THE LIBRARY
ST. MARY'S COLLEGE OF MARYLAND
ST. MARY'S CITY, MARYLAND 20686

Coralline Algae, A First Synthesis

Author
H.W. Johansen, Ph.D.
Associate Professor of Botany
Department of Biology
Clark University
Worcester, Massachusetts

CRC Press, Inc.
Boca Raton, Florida

Library of Congress Cataloging in Publication Data
Johansen, H William.
  Coralline algae, a first synthesis

   Bibliography: p.
   Includes index.
   1. Corallinaceae. I. Title.
QK569.C8J64   589.4'1   80-14028
ISBN 0-8493-5261-4

This book represents information obtained from authentic and highly regarded sources. Reprinted material is quoted with permission, and sources are indicated. A wide variety of references are listed. Every reasonable effort has been made to give reliable data and information, but the author and the publisher cannot assume responsibility for the validity of all materials or for the consequences of their use.

All rights reserved. This book, or any parts thereof, may not be reproduced in any form without written consent from the publisher.

Direct all inquiries to CRC Press, Inc., 2000 N.W. 24th Street, Boca Raton, Florida 33431.

© 1981 by CRC Press, Inc.

International Standard Book Number 0-8493-5261-4

Library of Congress Card Number 80-14028
Printed in the United States

PREFACE

As the title implies, this book is a first step at pulling together the voluminous but scattered information on coralline algae. Much can be said about these omnipresent plants of the sea, and the purpose here is to provide a coherent framework of data and discussion. This is the time to make this step, because research on coralline algae is now resulting in a surge of papers. What was once a rather neglected group of seaweeds is now becoming relatively well-known. Marine biologists are rapidly recognizing the important role that coralline algae play in the sea. And, the old idea that coralline algae are an extraordinarily difficult group to work with is also being dispelled.

More than most other seaweed groups, the Corallinaceae are well-defined. Hence, it is easy to scan papers and quickly pick out information on this group. There are two types of papers: those in which the research is focused on coralline algae and those in which data on these algae are part of a larger whole. A great deal of important literature was published from 1700 to the early 1900s. I have incorporated some of this, that which seems pertinent, especially in the chapter on taxonomy, but many others have been excluded. Rather, the aim of this book is to give a state-of-the-art presentation and to emphasize recent publications. Therefore, I have included as much information as possible up to and including 1979.

Another facet that reflects the well-defined status of the coralline family is the specialized vocabulary that has crept into use. I hope the glossary will help in this regard.

This book is aimed at marine biologists in general, and I have tried to keep the writing such that nonphycologists will not have undue difficulty with it. It is true, however, that some chapters, for example those on interal structures and reproduction, will be most understandable by phycologists.

In selecting chapter topics I kept in mind coralline algae— the plants. At the same time I recognize that of the wider interest to marine biologists is the role that these plants play in the seas. Hence, the 12 chapters form 3 groups: (1) plant structure and organization, (2) ecology, and (3) taxonomy and phylogeny. Hopefully, there is an approximately even distribution in these three areas. Coverage in the area of physiology is lacking, however. The closest this book comes to physiology is in Chapter 6, on calcification — a must with coralline algae.

Many people have helped with this book, both in reading for content and grammar, and in logistics, such as typing, drawing, and procuring publications. I thank the following people for reading parts of the manuscript or for exchanging ideas about coralline algae with me: W. H. Adey, V. Ahmadjian, H. Akioka, W. Andrake, L. F. Austin, E. A. Boger, M. A. Borowitzka, J. J. Brink, J. Cabioch, Y. M. Chamberlain, M. Chihara, B. J. Colthart, G. F. Elliot, D. J. Garbary, B. T. Gittins, M. H. Hommersand, S. E. Johnson, L. Irvine, P. A. Lebednik, M. Lemoine, M. M. Littler, T. Masaki, P. J. Matty, R. E. Meslin, J. D. Milliman, J. N. Norris, R. F. Nunnemacher, R. B. Searles, P. C. Silva, R. A. Townsend and W. J. Woelkerling. And there are surely others that I have inadvertently omitted. In spite of their help, it is possible that there are errors in the text, and for these I am fully responsible.

Many thanks go to Elizabeth M. Rogers and Inis C. Cook, as well as Terry Reynolds, Rene Baril, Teri McCall, and Roxanne Rawson of the Secretarial Pool, Clark University, for hours of tedious typing. Thanks also go to R. B. Parker, R. E. Levenbaum, Irene W. Walch, and Marion Henderson for help with illustrations and references. The Cooperation of CRC Press is much appreciated: Lisa Levine Eggenberger, Benita Budd Segraves, and B. J. Starkoff. The initial days of work on this book in a lovely Montana cabin were possible because of Sylvia and Neil R. Schroeder, and S. S. Cook, Jr. (Bud).

I would like to express my sincere gratitude to G. F. Papenfuss for starting me off on my study of coralline algae and for his continued interest. Much credit also goes to my family, Eric J., Brian F., Edith L., and Fredrik Johansen, as well as to Frances L. Pedusey, Carolee A. Virgilio, and Barbara J. Johansen for their understanding and support during the years when I was devoted to studying plants of the sea.

H.W. Johansen

THE AUTHOR

H. William Johansen, Ph. D., is Associate Professor of Botany, Clark University, Worcester, MA. He obtained a B. A. degree from San Jose State University, an M.A. Degree from San Franciso State University in 1961, and a Ph.D. degree from the University of California in Berkeley in 1966. From 1966 to 1968 he had an N.I.H. postdoctoral fellowship, part of which was spent in South Africa and Europe, and part in Berkeley. From 1968 to the present, he has held a position at Clark University in the Department of Biology.

Dr. Johansen has been actively involved in teaching plant and marine-oriented courses. His research deals with coralline algae, particularly systematics and ecology. Recently he has traveled to the Gulf of California, Japan, and Canada to study these algae.

He is a member of the American Association for the Advancement of Science, the International Phycological Society, the American Phycological Society, the British Phycological Society, Sigma Xi, and the Western Society of Naturalists.

Since his career in research began, Dr. Johansen has published more than 30 papers and abstracts, mostly on coralline algae. *Coralline Algae, A First Synthesis* is his first book.

This book is dedicated to
    my parents, Edith and Fredrik and
    my sons, Eric and the late Brian.

TABLE OF CONTENTS

Chapter 1
Scope and Diversity 1
Introduction 1
Basic Structure 1
Role in the Oceans 8
Classification 10
Summary 10

Chapter 2
Vegetative Cytology 13
Introduction 13
Cell Walls 13
Primary Pit-Connections 16
Secondary Pit-Connections 16
Cell Fusions 19
Nuclei 20
Organelles and Inclusions 20
Spore Germination 23
Sporeling Growth and Development 25
Meristems 27
Epithallia 29
Trichocytes and Megacells 33
Summary 36

Chapter 3
Structure of Nonarticulated Coralline Algae 39
Introduction 39
Lithophylloideae 41
Lithophyllum Series 42
Dermatolithon Series 42
Mastophoroideae and Melobesioideae 44
Thin Crusts 44
Ribbon Corallines 46
Thick Crusts 47
Unattached Coralline Algae 48
Epiphytic Coralline Algae 48
Parasitic Coralline Algae 52
Summary 55

Chapter 4
Structure of Articulated Coralline Algae 57
Introduction 57
Corallinoideae 57
Intergenicula 57
Genicula 60
Branching 63
Frond Initiation 65
Amphiroideae 66
Amphiroa 67

Genicula...68
Lithothrix..72
Metagoniolithoideae......................................74
Summary..76

Chapter 5
Reproduction..79
Introduction...79
Vegetative Reproduction..................................79
Conceptacles...79
Tetrasporangial Conceptacles.............................84
Tetrasporangia and Bisporangia...........................91
Male Conceptacles..92
Spermatangia...97
Carpogonia...98
Carposporophytes..101
Life Histories..106
Summary...108

Chapter 6
Calcification..111
Introduction..111
Cell Wall Composition...................................111
The Calcification Process...............................112
Summary...117

Chapter 7
Phytogeography.......................................119
Introduction..119
Cold Northwestern Atlantic..............................119
Cold Northeastern Atlantic..............................124
Cold Northwestern Pacific...............................127
Cold Northeastern Pacific...............................129
Tropical Regions..129
Below 30° South Latitude................................133
Summary...133

Chapter 8
Growth and Environment...............................135
Introduction..135
Spores and Substrates...................................135
Seasonality...136
Temperature...137
Light...138
Desiccation...139
Water Motion..140
Seawater Ingredients....................................142
Biotic Interactions.....................................143
Summary...147

Chapter 9
Production...149

Introduction 149
Growth Rates 149
Colonization and Succession 150
Organic Production 151
Inorganic Production 152
Unattached Coralline Algae 153
Fate of the Calcite 154
Summary 157

Chapter 10
Reef Building 159
Introduction 159
Basic Reef Structure 159
Grazing and Coralline Development 161
Indo-Pacific Reefs 162
Porolithon in the Pacific 163
Caribbean Reefs 164
Coralline Ridge-Formers in the Caribbean 168
Summary 170

Chapter 11
Fossil Coralline Algae 173
Introduction 173
Solenoporaceae 173
Ancestral Coralline Algae 175
Appearance of the Corallinaceae 176
Summary 177

Chapter 12
Taxonomy 179
Introduction 179
History 179
Recognizing the Genera 183
Current Schemes of Classification 187
Adaptations in Coralline Algae 188
Summary 191

References 193
Glossary 209
Appendix 1 215
Appendix 2 225

Index 227

Chapter 1

SCOPE AND DIVERSITY

INTRODUCTION

Seaweed evokes visions of waving blades of olive drab, not rockhard crusts or segmented branches in pinks and purples—living plants so hard they cannot be dented by a fingernail. How many of us would suspect that evolution could have led to such unlikely plants as coralline algae? Yet that is what swimmers see in coastal waters the world over: lime-impregnated seaweeds—the corals of the plant world. A great amount of time has allowed for the development of a diverse array of forms, habitats, and life styles. And that is what this book is all about—details of the anatomy, reproduction, and lives of coralline algae. These are plants, just as much as are lilies and palm trees. But these marine algae and terrestrial plants have been evolving separately for so long that about the only similarity is autotrophy. What do we suspect happened during these millions of years that led to hard seaweeds in the form of crusts, or segmented branches, or parasites, or gravelly nodules that roll about on the ocean floor? We can only speculate, but these calculated guesses are based on hundreds of studies of extant species, and some of extinct species. Still, they are plants that few people know about and some do not dream exist.

Coralline algae are pink, calcified red algae living in most euphotic zones where stable surfaces are present. They occur mostly as inflexible crusts or branched, articulated fronds, whereas elsewhere they occur as free coralline nodules, sometimes in great masses. These plants belong to the Corallinaceae, a large family in the red algal order Cryptonemiales. They are united in this family for several reasons, the most prominent being the copious amounts of calcium carbonate in their cell walls. Coralline algae are the hardest plants; they are the corals of the plant world. Noncorallinaceous marine algae, such as species of *Galaxaura, Padina, Peyssonellia,* and *Halimeda,* also calcify, but they do not become as hard as coralline algae. Dipping coralline algae into acid dissolves the calcium carbonate and then the plants are as soft as other red algae. Some coralline algae live with coral animals and are prominent in building tropical reefs. However, with the hardness and the reef-building characteristic the resemblance to corals ends.

In addition to their ability to calcify, coralline algae have other distinctive features, such as reproductive cells in roofed conceptacles, small vegetative cells organized into filaments, and an epidermis-like covering over most calcified surfaces.

There are enormous numbers of published reports dealing with coralline algae. Many of these papers have been incorporated into several review articles which have appeared lately: Littler[284] Adey and MacIntyre,[24] and Masaki[303] on nonarticulated species, Johansen[231] on articulated species, and Cabioch,[75,76] Bressan,[56] and Johansen[232] including information on both types. On the following pages emphasis will be on a state-of-the-art presentation.

BASIC STRUCTURE

For convenience, and also because it reflects phylogenetic differences, the coralline algae have customarily been divided into two groups, the nonarticulated and the articulated species. An idea of the variety of form in the two groups can be obtained from Figures 1 and 2 and Tables 1 and 2. In the nonarticulated group are thin (less that 200 µm thick) or thick crusts which grow slowly over a substrate. Some of these crusts

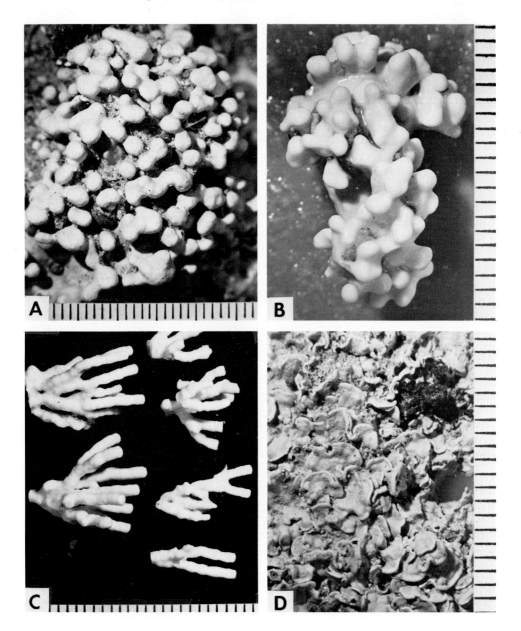

FIGURE 1. Nonarticulated coralline algae. A. *Lithophyllum*, an epilithic specimen. B. *Lithophyllum*, a specimen found lying free on the sea bed. C. *Neogoniolithon*, a specimen producing long protuberances which were broken from the base of the plant when collected. D. *Porolithon sonorense*, a species forming plate-like lobes. All scales in millimeters. (Fig. 1C, Johansen, H. W., *Phycologia*, 15(2), 221, 1976. With permission.)

have knobby protuberances that may be low or as much as 2 cm high. Protuberances may break off and continue to grow as free-living, gravel-like plants called *marl*. Other coralline algae grow concentrically around nuclei so as to form subspherical balls known as *rhodoliths*. Some nonarticulated entities are small specialized parasites growing on other coralline algae.

Articulated coralline algae consist of branching flexible fronds attached to crustose or rhizome-like holdfasts. The fronds are made up of small calcified segments, called *intergenicula*, which are separated from one another by uncalcified nodes called *geni-*

FIGURE 2. Articulated coralline algae. A. *Metagoniolithon radiatum.* Scale = 2 cm. B. *Cheilosporum proliferum*, a species having intergenicula with variously expressed lobes. Scale in millimeters. C. *Amphiroa anceps*, in which the intergenicula are flat. Scale in millimeters. (Fig. 2A, Ducker, S. C., *Aust. J. Bot.*, 27, 67, 1979; Fig. 2C, Johansen, H. W., *J. Phycol.*, 5(2), 118, 1969. With permission.)

cula. The fronds in some entities consist of only one or two intergenicula, but in most they are made up of hundreds and may attain a length of 30 cm.

As in most red algae, the cells of coralline algae are joined in filaments which are aggregated into thalli. In vegetative tissues these filaments consist of cells that are 5 to 15 µm wide and usually somewhat longer, even as much as 0.5 mm long in genicular cells in some articulated species (Figure 3). The continuity of a filament may sometimes be seen for many cells of its length because of intensely staining primary pit-connections. In crusts the lowermost filaments are oriented parallel to the substrate and the apices are at the thallus margins; these filaments consitute hypothallial tissue, the *hypothallus* (Figure 3). Branching from the uppermost hypothallial filaments are other filaments which become oriented perpendicular to the substrate and end at the upper

Table 1
AN ASSORTMENT OF GROWTH FORMS IN THE NONARTICULATED CORALLINE ALGAE

Forms	Examples
Thin, smooth crusts	*Fosliella farinosa*
Thin crusts repeatedly overgrowing one another	*Lithoporella melobesioides*
Branched, ribbon-like crusts attached at one end	*Metamastophora flabellata*
Thick, smooth crusts	*Clathromorphum circumscriptum*
Thin, loosely overlapping crusts, margins free	*Mesophyllum lichenoides*
Thick knobby crusts	*Neogoniolithon strictum*
Unattached branched forms called marl	*Lithothamnium corallioides*
Epiphytic crusts of determinate vegetative growth	*Clathromorphum parcum*
Unpigmented parasites, vegetative system reduced	*Kvaleya epilaeve*
Pigmented parasites, vegetative system endophytic	*Choreonema thuretii*
Pigmented, endophytic between cell wall layers in *Cladophora*	*Schmitziella endophloea*

surface of the crust. These filaments constitute perithallial tissue, the *perithallus*. Perithallial filaments are absent in some thin crusts, but in thick crusts these filaments may be many cells long. Plastids are generally present in perithallial cells and absent in hypothallial cells. Each perithallial filament terminates in one or several short, specialized cells which, in aggregate, form an epidermis-like layer called *epithallus*.

In the rigid protuberances in some crustose coralline algae, or the flexible fronds in articulated taxa, the terms hypothallus and perithallus are replaced by *medulla* and *cortex*, respectively (Figure 3). A medulla consists of a core of filaments extending through the branch. It is surrounded on all sides by cortical filaments which branch from the periphery of the medulla. Like perithallial tissue, cortical tissue is covered by epithallial cells. Although most medullary cells in articulated coralline algae are of the usual calcified type, uncalcified genicular cells occur at regular intervals. In some species these cells are much longer than the intergenicular medullary cells, although they are about the same diameter.

Thallus size increases by growth at the margins and surfaces of crusts and at the apices and intergenicular surfaces in branches. In most margins and apices epithallial cells are lacking and the meristematic cells terminate the filaments. These are called *primary meristems*. In calcified surfaces and in some margins and apices the meristematic cells are below epithallial cells and they constitute *intercalary meristems* (Figure 4).

Table 2
AN ASSORTMENT OF GROWTH FORMS IN THE ARTICULATED CORALLINE ALGAE

Forms	Examples
Extensive crustose base, fronds of 1 to 2 small intergenicula	*Yamadaea melobesioides*
Extensive crustose base, fronds of a few small intergenicula appressed to the holdfast	*Chiharaea bodegensis*
Fronds small, dense, pinnately branched sprays, conceptacles on intergenicular surfaces	*Bossiella plumosa*
Fronds robust, intergenicula flat	*Calliarthron cheilosporiodes*
Fronds delicate, intergenicula thin and terete	*Jania capillacea*
Fronds pinnately branched, conceptacles usually at branch apices	*Corallina officinalis*
Fronds dichotomously branched, intergenicula long and terete, conceptacles on intergenicular surfaces	*Amphiroa zonata*
Fronds, dorsiventrally oriented, intergenicula large, flat, irregularly shaped	*Bossiella californica* ssp. *schmittii*
Fronds dichotomously branched, intergenicula with pronounced upwardly pointing lobes	*Cheilosporum sagittatum*
Fronds robust, groups of branches arising from genicula, intergenicula terete, up to 3 cm long	*Metagoniolithon radiatum*

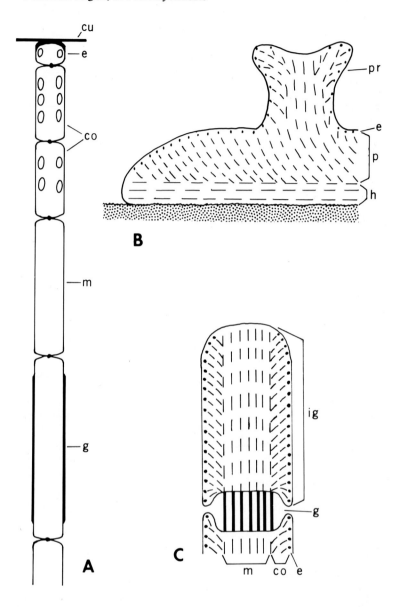

FIGURE 3. Diagrams of vegetative tissue. A. A filament showing, from top to bottom, the cuticle (cu), an epithallial cell (e), two cortical cells (co), a medullary cell (m), and an uncalcified genicular cell (g). B. A section through the margin of a crust and a protuberance (pr). Hypothallial filaments (h) are oriented parallel to the substrate (shaded) and perithallial filaments (p) arch upward and end in small epithallial cells indicated by dots (e). C. Apex of an articulated branch showing an intergeniculum (ig) and a geniculum (g). Medullary filaments (m) extend through the center of the branch. They are surrounded by cortical filaments (co), each of which ends in an epithallial cell (e).

In almost all coralline algae, reproductive cells are contained within conceptacles on the surfaces of crusts or intergenicula, or at the apices of the latter (Figure 5,6). In all conceptacles, except tetrasporangial ones in the subfamily Melobesioideae, a single pore is provided for the exit of spores and spermatia. Conceptacles are of three types: asexual, male, or female. Asexual conceptacles contain tetrasporangia (or bisporangia), male conceptacles spermatangia, and female conceptacles carpogonia. Following

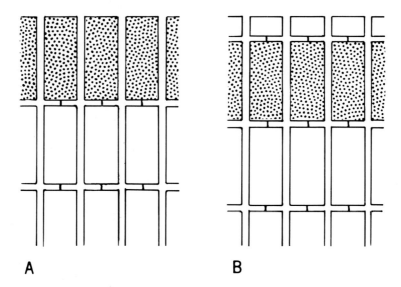

FIGURE 4. Diagrams of the two types of meristems in coralline algae. The meristematic cells are shaded. A. Primary meristem. B. Intercalary meristem.

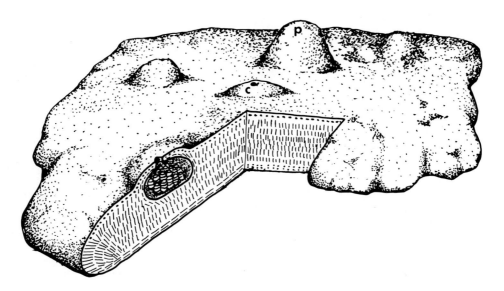

FIGURE 5. Drawing of part of a thick crustose coralline algae. c = conceptacle; p = protuberance. Drawn by W. Andrake.

the fertilization of carpogonia, development results in a small parasitic plant, a carposporophyte, within the female conceptacle. Carposporophytes produce carpospores.

Coralline algae have the *Polysiphonia* type of life history,[136] in that tetrasporophytes and gametophytes are similar in general appearance (Figure 7). More often than in most red algae, bisporangia instead of tetrasporangia are produced. These bisporangia usually contain two nuclei (one per spore) which are probably the same ploidy level as nuclei in the rest of the plant. This is not always the case, and binucleate bispores produced following meiosis probably give rise to haploid gametophytes. Normally, tetraspores germinate and give rise to gametophytes and carpospores give rise to tetrasporophytes.

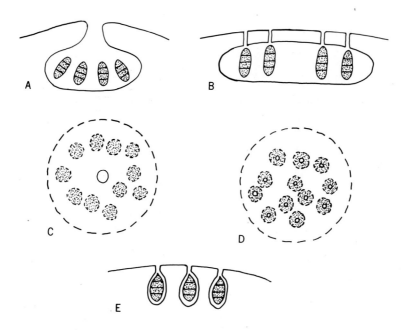

FIGURE 6. Diagrams of conceptacles. A. Section of single-pored conceptacle containing tetrasporangia. B. Section of multipored conceptacle with a pore above each sporangium. C. View from above of single-pored conceptacle with tetrasporangia beneath the roof. Outline of the chamber shown by large dashed circle. D. View from above of multipored conceptacle with one tetrasporangium beneath each pore. E. Section of unisporangial conceptacles in which a pore forms above each tetrasporangium. These are related to the conceptable in (B), except that the filaments among the sporangia have not broken down to form a chamber.

ROLE IN THE OCEANS

Coralline algae are present in almost all areas inhabited by seaweeds and often give a deep pink coloration to the habitat. Thus, unlike coral animals, they are common in cold waters as well as in warm waters. A few genera are restricted in their distribution, but most are widespread, although they usually tend to favor particular oceanic regions, apparently a factor of seawater temperature.

Features of the environment that are most significant in determining the presence of particular coralline species are largely unknown. In most species spores germinate best on rough substrates, although many taxa are epiphytic and some are even obligate parasites. Several environmental features foster the successful growth of coralline algae. One is low light intensity although, as for most of the other requirements, there are exceptions and a few species thrive in high light. In some areas coralline algae are the deepest algae, needing little light to survive and grow. Another feature of coralline algae is their inability to withstand drying, although a few exceptions live successfully in intertidal areas, provided moisture is retained on the thalli. Most species have a requirement for water motion. In fact, some communities dominated by coralline algae outcompete other plants and animals in areas of high wave energy, such as on the tops of tropical reefs. The chemical characteristics of seawater as they affect coralline algae are also largely unknown, but research suggests that some species grow well where domestic sewage in prevalent and other species have an uncommon sensitivity to phosphate.

Coralline plants are notoriously slow-growing. They grow by marginal extension over a substrate, by increasing thickness, and by branch elongation. They compete

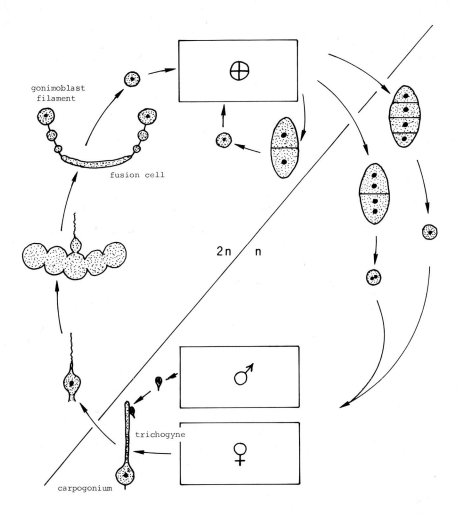

FIGURE 7. Life cycle in coralline algae. Asexual plants depicted by rectangle at top usually produce tetraspores. Tetraspores germinate and grow into male and female gametophytes, shown by the rectangles at the bottom. Bispores are sometimes produced, and if these are binucleated they presumably germinate and grow into sexual plants. If they are uninucleated they probably grow into asexual plants. Fertilization of carpogonia results in diploid carposporophytes (at the left) consisting of fusion cells producing gonimoblast filaments. The carpospores germinate and grow into asexual plants, and the cycle is complete. Critical nuclear events are meiosis in young tetrasporangia and in young bisporangia destined to produce binucleate bispores, and karyogamy in carpogonia.

successfully with other marine organisms, especially in upper subtidal areas where both nonarticulated and articulated forms often dominate climax communities. This prevalence accounts for much production of organic and inorganic materials, especially calcium carbonate. The carbonate produced may become part of a solid bioherm, or of sediment. The physiology of coralline biomineralization has been studied, but the details are poorly known.

Numerous observations of coralline algae have shown that on crowded shores they often form associations with other plants or animals. They serve as substrates for many animals and they themselves frequently grow on other marine plants or animals. Three coralline genera are parasitic on other coralline algae. Man has made little direct use of coralline algae, although underwater mining is carried out to obtain nodules for the liming of agricultural lands.

CLASSIFICATION

The classification of coralline algae has a long and involved history, as will be explored in chapter 12. From the 17th to the 19th century they were considered to be coral animals with microscopic polyps. Hense these plants were called zoophytes by Linnaeus[280] when he first named *Corallina officinalis* and a few other species. By the mid 1800s they were treated as red algae and several genera had been erected, for example, *Jania, Melobesia, Amphiroa, Arthrocardia, Lithothamnium* and *Lithophyllum*. During the latter half of the 19th century they were all placed in two groups, the nonarticulated (tribe Melobesieae) and the articulated (tribe Corallineae).[32] In the early part of the present century the nonarticulated forms were divided into seven groups,[161a,428] but the articulated forms were not subdivided until just a few years ago.[223] At the present time there are two main schemes for grouping the Corallinaceae. In this book the one that is used includes the nonarticulated genera in four subfamilies and the articulated genera in three.

From the time when *Corallina* was first described[280] more than 70 genera have been characterized. Several of them have been synonomized with previously described genera or eliminated because of nomeclatural irregularities, hence 43 are included in Tables 3 and 4.

Coralline algae are those seaweeds belonging to the Corallinaceae, a family in the order Cryptonemiales, class Florideophyceae and phylum Rhodophyta. The distinction between nonarticulated and articulated coralline algae is a useful one, and the presence or absence of genicula is phylogenetically basic. However, three lines of evolution probably lead from nonarticulated ancestors to articulated types. Hence, the articulated genera are placed in three subfamilies: the Corallinoideae, Amphiroideae, and Metagoniolithoideae. The four subfamilies of nonarticulated coralline algae are: Lithophylloideae, Mastophoroideae, Melobesioideae and Schmitzielloideae.

As may be expected with hard organisms, numerous fossils have been described. The complex of plants recognized as the Corallinaceae probably evolved from other calcified forms placed in the extinct family Solenoporaceae. The rich fossil record reveals that the Solenoporaceae existed at least until the Paleocene, whereas the first undoubted Corallinaceae has been recorded from the Jurassic.

Numerous characteristics have been employed in the various strategies for classifying the Corallinaceae on the generic level, and Johansen[232] has listed 41 of them. Features on which suprageneric taxa are based are developmental and structural, with the following structures used most often: (1) genicula, (2) tetrasporangial conceptacles, (3) intercellular connections, (4) vegetative tissues, (5) male conceptacles, (6) carposporophytes, and (7) sporelings. The subfamilies and genera are given in Tables 3 and 4.

SUMMARY

This book deals with the coralline algae, a group of red seaweed in the family Corallinaceae. The ability to deposit calcium carbonate in the cell walls characterizes most members of this family, and represents a condition analogous to that in coral animals. They are crusts or flexible fronds, or they may be unattached nodules. A great deal of literature has accumulated on these plants and in this book it is assembled and summarized in a state-of-the-art presentation.

Most coralline algae are thick or thin crusts, with or without protuberances, whereas others are articulated and consist of branched fronds made up of calcified intergenicula and uncalcified genicula. Some species are present as irregularly shaped nodules on the ocean floor. The basis unit of vegetative structure is the filament, many of which are united into thalli. Growth is by meristematic cells at filament apices or just below

Table 3
SUBFAMILIES AND GENERA OF NONARTICULATED CORALLINACEAE

Subfamilies and Main Features	Genera	
Lithophylloideae Direct secondary pit-connections present between cells; tetrasporangial pore plugs absent, i.e., conceptacles uniporate	*Dermatolithon* *Ezo* *Goniolithon*	*Lithophyllum* *Metamastophora* *Tenarea*
Mastophoroideae Cell joined laterally by fusions, no direct secondary pit-connections; tetrasporangial pore plugs absent, i.e., conceptacles uniporate	*Choreonema* *Lithoporella* *Fosliella* *Mastophora* *Heteroderma* *Hydrolithon* *Neogoniolithon* *Litholepis*	*Porolithon*
Melobesioideae Cells joined laterally by fusions, no direct secondary pit-connections; tetrasporangial pore plugs present and hence a pore forming above each sporangium	*Antarcticophyllum* *Chaetolithon* *Mastophoropsis* *Clathromorphum* *Melobesia* *Kvaley* *Mesophyllum* *Leptophytum* *Phymatolithon* *Lithothamnium*	*Sporolithon* *Synarthrophyton*
Schmitzielloideae Specialized endophyte in algal cell walls (*Cladophora*); uncalcified filaments lacking epithallial cells; no conceptacles, reproductive nemathecia erupting through cell walls of host	*Schmitziella*	

the terminal cells. The tissues of a crust consist of hypothallus and perithallus and a frond of medulla and cortex. Almost all calcified tissue is covered by a small-celled epithallus. Reproductive cells are produced within conceptacles which open to the sea by one or several pores. The life cycle is the common florideophycean type in which tetrasporophytes resemble gametophytes. Bispores instead of tetraspores are produced in some species.

Coralline algae grow slowly, but their large numbers make them important in the production of organic and inorganic material in the oceans. In most species the spores germinate on hard substrates, but a substantial number are epiphytic or parasitic. Most species grow best under conditions of low light intensity, are intolerant of desiccation, require water motion, and respond either negatively or positively to organic and inorganic ingredients in the ambient seawater. They occur in many temperature regimes throughout shallow waters. Dense populations form climax communities except in disturbed areas. Some species that thrive in high light conditions and turbulent water contribute to the building of tropical reefs. The fact that coralline algae have been in the seas for millions of years has allowed time for numerous interbiotic relationships to develop, some of them very intimate, such as in parasitic species.

Table 4
SUBFAMILIES AND GENERA OF ARTICULATED CORALLINACEAE

Subfamilies and Main Features

Amphiroidea
    Genicula one tier or, more often, several tiers of cells; direct secondary pit-connections present between cells; conceptacles on lateral surfaces of intergenicula

Genera

Amphiroa
Lithothrix

Corallinoideae
    Genicula single tier of long cells; cells joined laterally by fusions, no direct secondary pit-connections present

Alatocladia *Corallina*
Arthrocardia *Haliptilon*
Bossiella *Jania*
Calliarthron *Marginisporum*
Cheilosporum *Serraticardia*
Chiharaea *Yamadaea*

Metagoniolithoideae
    Genicula of many layers of cells, and producing branches; cells joined laterally by fusions, no direct secondary pit-connections present; conceptacles on lateral surfaces of intergenicula

Metagoniolithon

The family Corallinaceae is placed in the order Cryptonemiales, class Florideophyceae and phylum Rhodophyta. Numerous useful characteristics enable phycologists to group coralline algae into 7 subfamilies containing about 43 genera. Four of these subfamilies contain nonarticulated species and three of them articulated species.

Chapter 2

VEGETATIVE CYTOLOGY

INTRODUCTION

Although most coralline cells are encased in calcite-impregnated walls, the apparent lack of disruption when decalcified in acids has led to numerous studies at the light microscope level. In the 1970s several reports at the electron microscope level appeared. Foremost among recent cytological studies of coralline algae are those by Cabioch,[75] Cabioch and Giraud,[78,79] Giraud and Cabioch,[185,186] Bailey and Bisalputra,[34] Magne,[293] Borowitzka,[46] Borowitzka and Vesk,[50a] and Duckett and Peel.[148]

In this chapter cytological aspects of vegetative cells are considered as well as the structure and coordinated activities of tissues. Reproductive cells are included when necessary to the understanding of cytological features not having a bearing on reproduction. Genicular cells are treated in Chapter Four.

CELL WALLS

A coralline vegetative cell, such as one from perithallial tissue, consists of a lumen bracketed at each end by primary pit-connections and surrounded by a three-layered cell wall in which calcite is deposited. Most examinations of such cells have been made on decalcified, sectioned, and stained material in which each cell is clearly outlined by a presumably uncalcified inner cell wall layer. Illustrations of coralline tissues made at the light microscope level usually show this inner wall layer because it stains more intensely than the thicker, calcified, middle cell wall layer (Figure 1). The outer wall layer, the thin middle lamella, or the "primary layer"[159] is midway between cell lumina and is also calcified.

It has been claimed that the inner cell wall layer is cellulosic,[75] but in the past there has been disagreement over whether or not this compound occurs in coralline algae. Negative results for cellulose were obtained by Baas-Becking and Galliher[33] whereas Naylor and Russell-Wells,[328] Ross,[365a] and Myers and Preston[327a] stated that it is present. More recently Bailey and Bisalputra[34] saw microfibrils in the inner cell layers in some articulated coralline algae. The presence or absence of cellulose still remains unresolved, however, and McCandless[312] regarded the evidence for cellulose in the Florideophyceae as a whole inconclusive.

The bulk of the calcium carbonate is in the thick middle cell wall layer (Figure 2). Good evidence for the presence of the mineral was obtained by Cabioch[75] using polarizing lenses to view freshly prepared material. The outermost walls of growing margins in thin crustose forms, such as *Fosliella* (Figure 3), are uncalcified, suggesting that calcification first takes place in lateral walls and then occurs later in the transverse walls as they form. Thus, marginal cells may continue to grow in length unhampered by calcite. Similarly, calcite is absent from settled spores until it is secreted at the first division in the newly formed walls of the two cells.[169] Cabioch[75] noted that the calcified transverse cell walls in perithallial tissue of *Lithophyllum incrustans* were unusually conspicuous and that the mineral was even present around rhizoids growing from germinating spores of *Amphiroa rigida*.

Calcite permeates a pectic material in the middle cell wall layer,[436a] but further study of the organic wall substances is needed. This is especially important in understanding the process of calcification; it seems to involve a close relationship between an organic matrix and calcite crystals[45] (see Chapter 6). In cells near branch apices in *Corallina*,

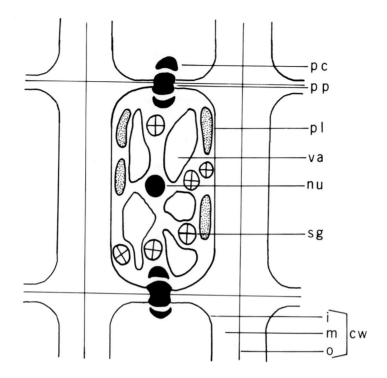

FIGURE 1. Diagram of a generalized coralline vegetative cell.

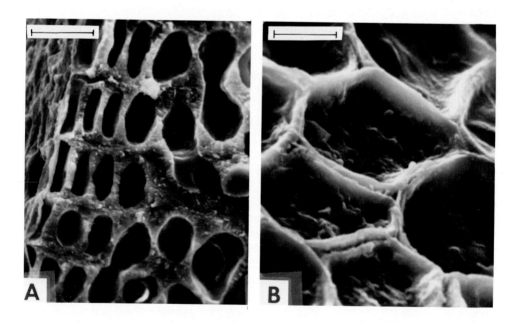

FIGURE 2. Cell walls in the crustose coralline *Phymatolithon calcareum*. A. Fractured specimen. Scale = 10 μm. B. Surface view of dried specimen showing walls and middle lamella. Scale = 4 μm. Both courtesy D. J. Garbary.

Calliarthron and *Bossiella*, Bailey and Bisalputra[34] saw vesicles 1 to 2 μm in diameter produced by Golgi bodies. These vesicles, and others 0.25 to 0.5 μ in diameter, were apparently discharging their contents into the cell wall layers when the plants were

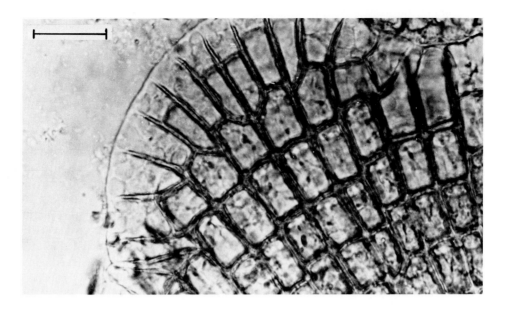

FIGURE 3. Living plant of *Fosliella zonalis* showing marginal meristem from above. Scale = 20 μm. Courtesy Y. M. Chamberlain.

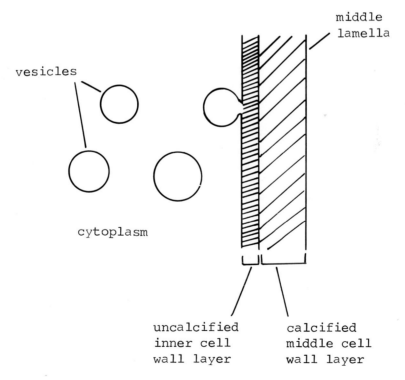

FIGURE 4. Diagram of the way vesicles may discharge material from the cytoplasm into the cell wall.[34]

fixed (Figure 4). This was the first time that vesicles of this type were reported in coralline algae, and additional studies are warranted.

The growth of calcite crystals occurs near the apices of medullary cells in *Corallina*.

The first signs of deposition are the appearance of small crystals about 250 nm long, mostly oriented with their long axes (c-axes) at right angles to the plasma membrane.[50] Farther from the membrane appear crystals that are larger and irregularly arranged and in the middle lamella they are approximately parallel to the cell surface.[159] Borowitzka and Vesk[50] described the fully formed crystals as thin rods up to 120 nm wide and about 8 nm thick.

Three species are exceptions to the rule that calcium carbonate is produced in coralline algae. The most notable are the highly modified *Schmitziella endophloea* and *S. cladophorae* which grow as pigmented filaments between the cell wall layers of *Cladophora*.[89,424] In fact, if these are coralline algae, then they have lost the metabolic ability to secrete calcite. This is also the case with *Melobesia van heurckii*, consisting of delicate crusts growing on hydroids.[75]

Under particular conditions calcified cell walls may lose their mineral impregnation, a process that is poorly understood. One of the most notable areas where decalcification occurs is in the formation of conceptacle primordia, a feature discussed later. In the vegetative system, the maturation of genicula in articulated coralline algae involves decalcification, although perhaps not in *Metagoniolithon*. Local peeling of patches of epithallial and perithallial tissues in some crustose corallines must involve decalcification.[75] Possibly, growth at crust margins and branch apices includes decalcification of certain cells when secondary tissues are generated from preexisting calcified tissues. There is a marked paucity of information on these aspects of coralline growth and development.

PRIMARY PIT-CONNECTIONS

Three types of cellular connections occur in vegetative tissues of coralline algae, namely, *primary pit-connections*, direct *secondary pit-connections* and *fusions* between laterally contiguous cells. Primary pit-connections form in all vegetative tissue as meristematic cells divide. During cytokinesis an infurrowing cell wall does not form a complete closure. Instead, as in other red algae, a dense *pit plug* forms to close the open space and hence the two protoplasts become separated (Figure 5). In medullary tissue of most articulated coralline algae and in certain crustose hypothallia which are coaxial, such as in *Mesophyllum*, cell divisions are synchronous and arching lines of primary pit-connections may be seen in the tissues.

Adey[3] (1964) described perithallial pit-connections in which the plugs (as plates) were bracketed by pit bodies staining intensely with phosphotungstic hematoxylin. Similar pit bodies occur in *Lithothamnium* and in *Leptophytum*[8] but not in the closely related *Clathromorphum*. Ultrastructural evidence presented by Borowitzka and Vesk[50] corroborates Adey's observations. Their photographs clearly show dense "caps" bracketing primary pit-connection plugs in *Corallina officinalis* and *Haliptilon cuvieri* (Figure 5). They found that the caps are usually present, but sometimes one or both are absent. Their study further confirmed that, except for the caps, pit-connections in coralline algae are similar to those in other red algae.

SECONDARY PIT-CONNECTIONS

Numerous workers, notably Rosenvinge[365] Suneson,[421,423] and Cabioch[72,75] have reported secondary pit-connections between laterally contiguous cels in coralline algae (Figure 6). They occur in two subfamilies, the Lithophylloideae and Amphiroideae and in scattered taxa of the Melobesiodeae. At the light microscope level they resemble primary pit-connections except for their location. In long celled tissues, for example medullary tissue in *Amphiroa* and hypothallial and perithallial tissues in *Dermatoli-*

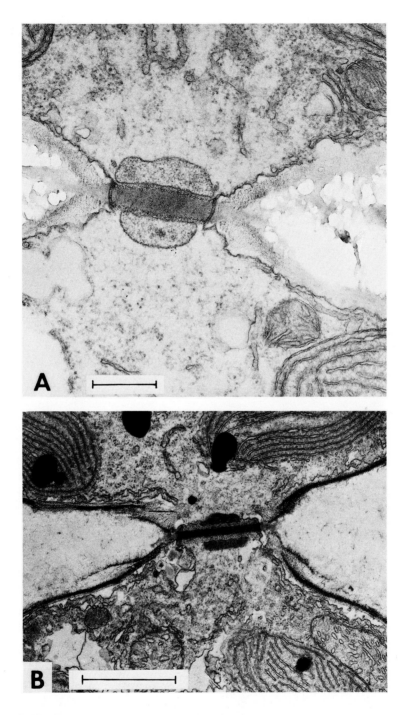

FIGURE 5. Primary pit-connections. Note that the pit plugs are bracketed by pit caps. A. *Corallina officinalis.* Scale = 0.5 μm. B. *Haliptilon cuvieri.* Scale = 1 μm. Courtesy M. A. Borowitzka and M. Vesk.

thon, secondary pit-connections tend to be located in the lateral walls in the upper halves of the cells. In shorter cells the locations seem to vary. Suneson[421] observed in coralline algae that a small cell was not budded off from an initiating cell prior to the formation of a secondary pit-connection (such as in the Rhodomelaceae). Therefore, he called these corallinaceous structures *direct* second pit-connections.

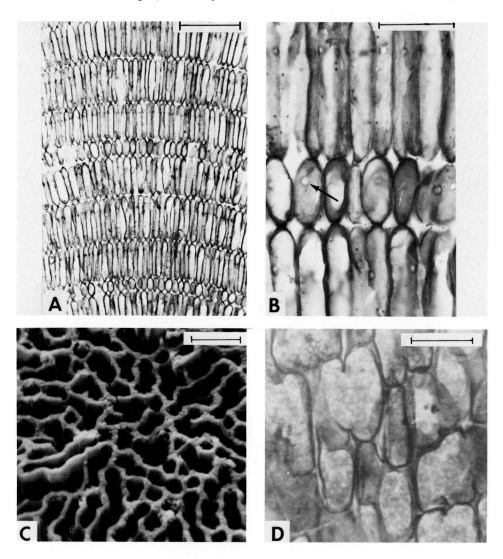

FIGURE 6. Vegetative tissues. A and B. Secondary pit-connections in *Amphiroa anceps*. A scale = 100 μm. B. Scale = 50 μm. (Fig. 6 A and B, Johansen, H. W., *Univ. Calif., Berkeley, Publ. Bot.*, 49, 1, 1969. With permission.) C. Cellular fusions in fractured intergeniculum of *Corallina*. Scale = 20 μm. Courtesy D. J. Garbary. D. Fusion in sectioned perithallus of *Neogoniolithon*. Scale = 20 μm. (Fig. 6D, Johansen, H. W., *Phycologia*, 15(2), 221, 1976. With permission.)

Cabioch[75] studied their development in *Lithophyllum tortuosum* and *Lithophyllum expansum* and concluded that there were two stages: (1) the growth of a narrow canal through the cell walls as they dissolved and (2) the formation of a pit plug (Figure 7). Thus, in meristematic cells tubules grow through the cell walls, making contact and fusing with neighboring cells. Sometimes a tube from one cell is met partway by a tube from another cell, after which they may or may not coalesce. If coalescence occurs, a partition soon forms to separate the two protoplasts. Whichever the case, a pit plug appears between the two cells and the pit-connection is completed.

Adey[7] suggested that in *Lithophyllum orbiculatum* secondary pit-connections formed in basal parts of some perithallial meristematic cells and, when divisions occurred, the new cells had preformed connections. He also reported rare secondary pit-connections between epithallial cells in this same species.

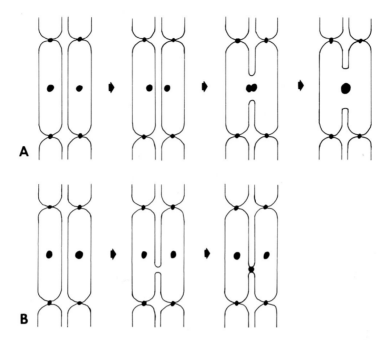

FIGURE 7. A. Diagrams showing stages in development of cell fusions. Note the interaction of nuclei (see text for details). B. The formation of a secondary pit-connection. A pit plug forms in the new pit-connection.

CELL FUSIONS

The tendency for cells in coralline vegetative tissue to unite by large canals has been shown by many anatomists (Figure 6). This has been reported in all subfamilies except the Lithophylloideae and Amphiroideae, although there are a few reported exceptions. Suneson[423] found occasional fusions in *Dermatolithon litorale* and *D. corallinae* near conceptacular pores, and Adey[7] and Masaki[303] found them in some species of *Lithophyllum*. The fusions have been seen in hypothallial, perithallial, medullary, and cortical tissues and in genicular tissue only in *Metagoniolithon*.[144] In most entities fusions (as well as secondary pit-connections) between epithallial cells are lacking, but Lebednik[262] noted that they were occasionally present in the thick epithallia of *Clathromorphum*.

Cabioch[75] presented observations on the development of fusions and the interplay of nuclei. In *Mesophyllum lichenoides* hypothallial cells begin to fuse when nuclei in two or more laterally adjoining cells migrate close to the wall where the fusion will occur. As the walls dissolve, the nuclei come to cluster in the area of dissolution (Figure 7). Breakdown continues until part or all of the lateral walls between two or more cells has disappeared. Eventually, the nuclei fuse and, in effect, a single cell that may be as much as 50 to 60 μm in diameter is formed. Fusions in *Phymatolithon lenormandii* were also studied,[75] especially in perithallial cells where usually only two cells are involved per fusion.

Except for the Amphiroideae, cell fusions also occur in articulated coralline algae. Cabioch[75] examined this phenomenon in *Corallina officinalis,* where fusions originate in newly formed medullary tissues and occasionally involve several cells. In these elongate cells, the close correspondence between nuclear location and the site of cell wall dissolution was clearly evident (Figure 7.) Again, nuclear fusions follow resorption of the cell walls. Sometimes more than half of the walls between two cells disappears.

The biology of cellular fusions and the concomitant nuclear fusions, something studied long ago by Rosenvinge,[365] needs to be examined using modern techniques. Fusions seem to embody a fundamental similarity among all coralline algae, except those in the two subfamilies Lithophylloideae and Amphiroideae. Cabioch[75] has shown that the formation of secondary pit-connections and cellular fusions may be compared. In direct secondary pit-connections there is no apparent involvement of nuclei, the initiating canals are narrow, and formation originates in meristematic tissue and is concluded rapidly. On the other hand, fusions involve cell wall dissolution between adjacent cells so as to form coenocytes, nuclear activity is a prerequisite, and these fusions occur in tissue that has already differentiated. The fact that a canal forms early in pit-connection formation points to the probability of a common origin for the two types of connection.

It has been stated that the subfamily Melobesioideae has cell fusions and lacks secondary pit-connections.[223] However, recent work has shown that this is not always true. In the primitive species *Sporolithon erythraeum (as Archaeolithothamnium)*, Cabioch[72] occasionally found cell fusions and secondary pit-connections in the same cell. Even more striking is the report by Townsend[435a] stating that cell fusions in the monotypic *Synarthrophyton* are confined to the roofs of carposporangial and tetrasporangial conceptacles whereas secondary pit-connections occur in both hypothallial and perithallial tissues. On the other hand, cell fusions have been reported in the Lithophylloideae, although rarely. These reports dim the usefulness of cellular connection characteristics in segregating subfamilies in the Corallinaceae.

NUCLEI

Many cytologically oriented papers include observations on nuclei, but still not a great deal can be said about them. Notable for careful work with the light microscope are Davis,[109] Westbrook,[445] Suneson,[426,427] Magne,[293] and Cabioch,[75] some of their papers dealing mostly with nuclei in reproductive cells.

Single nuclei are present in all coralline cells except in fused cells described earlier and in certain reproductive cells (e.g., carposporophytic fusion cells). Nuclear shape varies although it is mostly spherical or subspherical. Johansen,[223] reported lobed and elongate nuclei in genicula of *Calliarthron*. Nuclear size also varies considerably — 2 to 6 μm is the usual diameter range for vegetative cells, whereas tetrasporangia and carposporangia may have nuclei as large as 10 to 16 μm in diameter.

Magne[293] described and illustrated somatic mitosis in cortical meristems of *Corallina*, as did Westbrook.[445] Recorded and illustrated were the usual stages of mitosis, including the disappearance of the nucleolus and nuclear membrane. More information on nuclei is presented under tetrasporangia and spermatangia in the chapter on reproduction.

Chromosome complements have been determined in several coralline algae (Table 1), and the two basic haploid numbers are $n = 16$ and $n = 24$.[135] Of the eight species for which numbers are known, it is only in two species of *Dermatolithon* that the haploid numbers are 16.[427] *Marginisporum, Corallina, Jania, Metamastophora*, and *Fosliella* were all $n = 24$.

ORGANELLES AND INCLUSIONS

As seen in Table 2, several subcellular structures are present in vegetative cells, but information on them is scanty. Spherical and elongate plastids are particularly abundant in surface cells, decreasing in number and mass in deeper cells until hypothallial and medullary cells mostly lack them. Observations with the electron microscope have

Table 1
CHROMOSOME NUMBERS IN THE CORALLINCEAE[135]

Species	Numbers	Ref.
Marginisporum aberrans	n = 24 (spermatia)	379
Corallina officinalis	n = 24 (spermatia)	293, 421
Corallina elongata	n = 24	453
Jania rubens	n = 24 (spermatia)	421
Fosliella farinosa	n = 24 (meiosis in tetrasporangia)	35
Metamastophora lamourouxii	n = 24 (spermatia)	426
Dermatolithon corallinae	n = 16 2n = 32	427
Dermatolithon litorale	2n ≅ 30	427

Table 2
ORGANELLES AND OTHER CELLULAR INCLUSIONS IN VEGETATIVE CELLS OF CORALLINE ALGAE

Structures	Comments	Ref.
Nuclei	See text	75
Plastids	Typical of red algae	46, 50, 185, 468
Starch grains	See text	50, 50a, 75
Staining bodies	Possible proteinaceous storage, restricted to some species	3, 75
Crystals	In cytoplasm, only reported in *Phymatolithon lenormandii*	75
Golgi bodies	But also see Duckett and Peel[148] for carposporangia	186, 468

revealed that they are of the usual rhodophyte type (Figure 8), including phycobilisomes.[50,148,185,186] In an ultrastructural study of plastid development in carposporangia of *Lithothrix aspergillum* Borowitzka[46] described their transformation from proplastids. Stages of development include (1) the appearance of a peripheral thylakoid (the inner limiting disc),[148] the first membranous structure to appear within the plastid, (2) an irregular tubular membrane system (tubules 30 to 35 nm in diameter) in the DNA region of the plastid, the so-called pro-lamellar-like body, (3) more thylakoids appear, possibly originating from the pro-lamellar-like body, to which some of them are attached, (4) organization of the thylakoids into parallel groups, (5) the single DNA region fragmenting into several areas within the plastid, and (6) phycobilisomes originating on the outer thylakoid surfaces. Carposporangia are excellent models for studying plastid development; young carposporangia are small unpigmented cells whereas mature carposporangia are large red spheres. Duckett and Peel[148] also studied carpo-

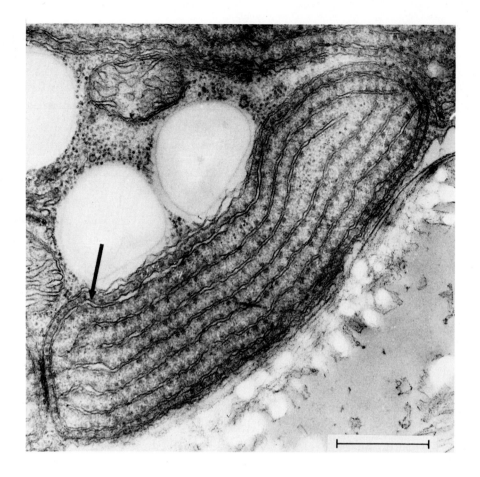

FIGURE 8. Plastid in vegetative cell of *Fosliella* sp. The peripheral thylakoid is marked by an arrow. Scale = 0.5 μm. Courtesy M. A. Borowitzka and M. Vesk.

sporangial plastid development, but in *Corallina officinalis* and *Jania rubens*, obtaining similar results.

Golgi bodies have been examined in developing carposporangia of *Corallina officinalis* and are unusual.[148] Instead of the usual gaps between successive cisternae, the narrowed spaces contain a dense electron opaque material. These types of Golgi bodies have also been observed in the sporangia of some noncorallinaceous red algae.

Vacuoles are small in peripheral thallus cells, but they become larger in deeper tissues. Hypothallial and medullary cells appear dead because of large vacuoles and sparse cytoplasm in which the most prominent structures are nuclei and starch grains.

Starch grains are common in coralline cells, particulary in internal tissues such as hypothallia and medullae. Borowitzka and Vesk[50] suggested that the general absence of starch grains in peripheral photosynthetic cells is an adaptation allowing for as much light as possible to strike the plastids in these cells. After all, dim habitats and dense cell walls already reduce the amount of light impinging on the plastids. As in other red algae, they develop external to the plastids.

According to Cabioch[75] starch grains vary in size from 0.5 μm in diameter in *Dermatolithon cystoseirae* to 10 to 15μm in diameter in *Lithophyllum expansum* as (*Pseudolithophyllum*). In her developmental study of starch grains in *L. expansum*, Cabioch described delicate net-like structures near nuclei in the marginal meristematic cells. As the cells became more deeply embedded in the thallus, small grains developed from this net and eventually grew to full size. The nucleus appears to play a role in starch grain formation.

Starch grains also develop within maturing carposporangia. Borowitzka[46] showed that in *Lithothrix aspergillum* the grains form in intimate association with endoplasmic reticulum, a feature that may occur also in other red algae. The endoplasmic reticulum was observed not to be in direct contact with plastids, and it may be that these membranes provide the enzymes for polymerization.

Other poorly understood inclusions are present in certain taxa of the Melobesioideae. Adey[3] reported the consistent presence of densely staining bodies as large as 5 μm in diameter in *Phymatolithon laevigatum;* they may represent protein storage. Their absence in *Phymatolithon rugulosum* renders them useful in distinguishing the two species. Cabioch[75] reported them also in *Lithothamnium sonderi.* Further, she described small (2 μm) cytoplasmic crystals in *Phymatolithon lenormandii.*

SPORE GERMINATION

Tetraspores, bispore, and carpospores are large, approximately spherical cells, although carpospores often have several slightly flattened surfaces resulting from mutual compression. Tetraspores are often discharged in groups of four still surrounded by the tetrasporangial wall; carpospores are usually free from one another after discharge. Both types of spores are amply endowed with red plastids, and a spore mass may be visible through thin-walled conceptacles, particularly if the plant has been decalcified.

Spore walls are delicate and uncalcified. Recent electron microscopy of carpospore walls by Duckett and Peel[148] revealed well-defined "particles" immediately outside the plasma membrane in *J. rubens,* a feature lacking in *C. officinalis.* The particles may be related to the attachment requirements of *J. rubens,* an epiphytic species. It also supports the taxonomic distinction between these two entities, at one time placed in the same genus. The smallest spores are those in the two subfamilies having secondary pit-connections, the Lithophylloideae and the Amphiroideae (Table 3). Bispores tend to be larger than tetraspores because both spore types are derived from sporangia of the same size. The consistently larger spore diameters in the Corallinoideae sets it apart from the Lithophyllideae and the Amphiroideae. Carpospores in some species of *Jania* are larger than in any other coralline algae, being more than 100 μm in diameter.

Because of the relatively great number of tetrasporangial plants, most studies of spore germination have been with tetraspores. Coralline algae having mature conceptacles readily discharge spores when placed in seawater. The spores sink, become attached to hard surfaces such as microscope slides, and begin to germinate, usually in a matter of hours.

It is convenient to break down the examination of sporeling development into (1) germination of spores into four-celled sporelings (Figure 9), (2) development of four-celled sporelings into 20- to 32-celled plants still contained within the original spore wall, and (3) the first stages of growth in thickness, the formation of discrete marginal meristems, and the subsequent expansion over the substrate. Although nearly all coralline algae exhibit the discoid *Dumontia* type of spore germination, a few parasitic and endophytic species have the filamentous *Naccaria* type,[76] a specialized form considered later.

Early stages in the germination of coralline spores have been described in many papers, the most recent ones by Chihara[94-100] and Notoya.[331-333] Spores sink quickly and adhere to a substrate by means of a mucilaginous envelope which varies in thickness. This envelope is relatively thick in *J. rubens,* where it is probably composed of mucopolysaccharides, and thin in *C. officinalis.*[241] In all coralline algae having the

Table 3
APPROXIMATE SPORE DIAMETERS IN SOME CORALLINE ALGAE

Taxa	Diameters (μm)	Ref.
Tetraspores Amphiroideae	20—40	464, 94
Bispores Amphiroideae	50—60	94
Tetraspores Corallinoideae	45—80	95, 96
Bispores Corallinoideae	90—100	94
Spores *Choreonema thuretii*	25—30	76
Tetraspores Lithophylloideae	20—40	98
Tetraspores (?) *Melobesia pacifica*	75—110	332,333
Tetraspores (?) *Heteroderma sargassi*	20—30	332,333
Carpospores *Jania pusilla*	100—130	147
Carpospores *Jania rubens*	70—90	147

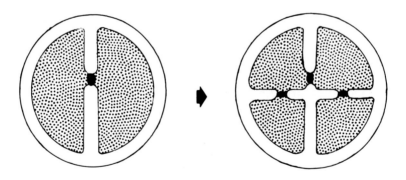

FIGURE 9. Diagrams of initial stages in spore germination showing primary pit-connections indicating cell lineages.[76]

Dumontia type of germination, development of the spores into four-celled sporelings is identical. An attached spore rapidly divides into four cruciately arranged cells by new cell walls which form perpendicular to the substrate (Figure 9). The first division yields two cells, each of which divides anticlinally with respect to the first-formed wall, the result being four cells all of about equal size and in contact with the substrate. Primary pit-connections are present from the time of the first division, and their positions enable division sequences to be followed by an investigator.[76] During these divisions and in the immediately subsequent ones, there is little increase in the mass of the sporeling. Cabioch[62] pointed out that each of the four, so-called primordial

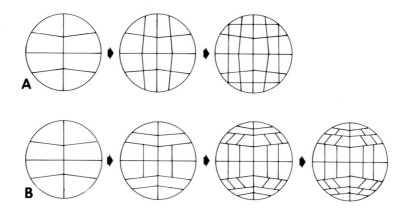

FIGURE 10. Patterns of cell division in young sporelings. A. *Corallina*. B. *Amphiroa*.[100]

cells is autonomous and able to develop into a coralline thallus even if the other cells should abort, a frequent occurrence in her study of *C. officinalis*.

SPORELING GROWTH AND DEVELOPMENT

Studies of development from four-celled sporelings into small plants that are no longer contained within the original spore walls reveal several patterns of cell division that correspond to taxonomic groupings. At least seven types have been described (Figures 10 and 11). Figure 10 shows the division leading to a 32-celled sporeling in *Corallina*[96] in which all the cells are in contact with substrate, and the plant is scarcely larger than the attached spore from which it grew. Numerous other species in the Corallinoideae and Melobesioideae, and some of the Mastophoroideae, have this pattern of cell division. Yamada[462] called this the *Corallina* type of spore germination.

Several species of *Amphiroa* exhibit the pattern shown in Figure 10.[94] This, the *Amphiroa* type[462] results in 20- to 32-celled sporelings that are clearly different from those in *Corallina*. Note how the walls in adjacent cells are arranged in relatively straight lines in the *Corallina* type in contrast to those in the *Amphiroa* type. The number of cells produced within the original spore wall varies somewhat depending on spore size. The divisions occur extremely rapidly in these sporelings and, in *Amphiroa zonata*, the plants may consist of 32 to 36 cells only 12 hr after spore attachment.[97] *Lithothrix aspergillum*[94,223] as well as several species in the Lithophylloideae[100,332] have the *Amphiroa* type of sporeling development.

In a summary of the six patterns then recognized, Chihara[100] listed, in addition to the *Corallina* and *Amphiroa* types, the *Fosliella minutula*, *Fosliella farinosa*, *Porolithon onkodes*, and *Fosliella farinosa* f. *solmsiana* types (Figure 11). Finally, Notoya[332] described another mode of cell division in *Melobesia pacifica* from Japan (Figure 11). This pattern is unique in that many of the divisions result in wedge-shaped cells and a marginal meristem that takes form earlier than in other species.

At varying times during the later stages of development, usually after the 16-celled stage has been reached, two changes take place. First, in all but the simplest species, such as *Fosliella* spp., the single layer of cells becomes multilayered when some of the interior cells cut off cells above. Second, a layer of cells completely or partly surrounding the sporeling is produced from the basal stratum of cells. This is a marginal meristem, and the basal stratum is a hypothallus. The formation of this meristem is generally rather sudden, but the plant now has the capability of increasing its surface coverage. Cabioch[76] showed how this meristem first appears and that there may be

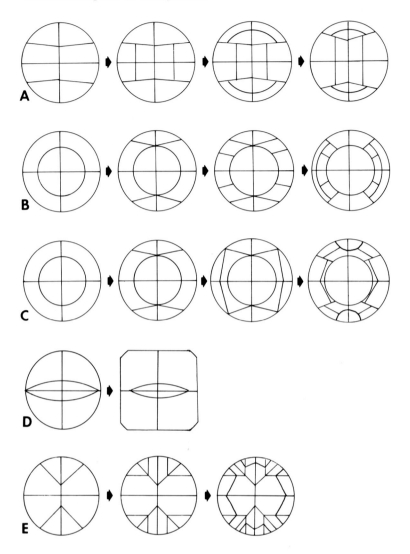

FIGURE 11. Patterns of cell division in young sporelings. A. *Foliella minutula*. B. *Fosliella farinosa*. C. *Porolithon onkodes*. D. *Fosliella farinosa* f. *solmsiana*.[100] E. *Melobesia pacifica*.[332]

several strata of marginal meristematic cells in older sporelings (e.g., in *Phymatolithon lenormandii*). These meristematic cells are unpigmented, in contrast to the more central cells in the young crust.

After observing transformations from spore to young crust in many taxa, Cabioch[76] suggested that there are three modes of development in the coralline algae forming *Dumontia* type crusts: (1) "Le mode *Lithothamnium*," (2) "Le mode *Neogoniolithon*," and (3) "Le mode *Lithophyllum*." Her assignments are based on differences that prevail until the crusts contain many cells instead of only about 32 as in the other categorizations described so far. The essential distinctions among the three types are:

1. The *Lithothamnium* mode of development involves the formation of several marginal meristems superimposed horizontally and the apparently irregular production of cells. Numerous species in the Melobesioideae and Corallinoideae fall in this category.

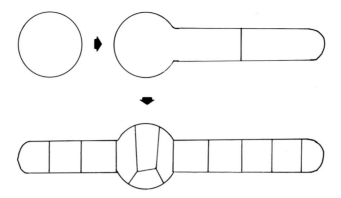

FIGURE 12. Diagram of the filamentous *Naccaria* type of spore germination in *Choreonema thuretii*.[76]

2. The *Neogoniolithon* mode entails a more regular arrangement of cells in the young crusts and the early formation of trichocytes. Included in this group are *Neogoniolithon notarisii* and some species of *Fosliella*.
3. The *Lithophyllum* mode is characterized by (a) a regular and simple cutting off of cells, (b) a single marginal meristem and (c) the early appearance of trichocytes and epithallial cells. As would be expected, secondary pit-connections appear early in development. Exhibiting this type of development are species of *Lithophyllum, Dermatolithon, Pseudolithophyllum,* and, perhaps, *Goniolithon*.

Surprisingly, the spores of two species of *Amphiroa* from the Mediterranean, *A. verruculosa* and *A. rigida,* germinate by the formation of filamentous protonema rather than discs.[69] Hence, these plants have a *Naccaria* type of germination.[463] Later Cabioch[76] germinated spores of the parasitic *Choreonema thuretii* and found that this germination was also of the *Naccaria* type (Figure 12). She suggested that the spores of *Schmitziella endophloea* germinate in the same manner. In *C. thuretii* the relatively small (Table 3) unpigmented spores extrude simple germination tubes which probably infect the host plants (species of *Jania, Haliptilon,* and *Cheilosporum*) in an unknown manner. In the two species of *Amphiroa* the growth of septated protonema is accompanied by cell divisions of the spores themselves and, in a sense, both the discoid and filamentous types of germination are exhibited.[69] Because the early growth of *Clathromorphum parcum* appears more or less similar to that in the species of *Amphiroa*,[69,76] it is possible that spore germination may involve a filamentous stage, although probably of limited extent.[23]

MERISTEMS

At the end of each vegetative filament, there exists a meristematic cell in terminal or subterminal position. Meristematic cells are distinguishable not only by their position, but also by their generally larger size, large nuclei, and more or less vacuolated but dense, cytoplasm (Figure 13). Ultrastructural features of apical meristematic cells in *Lithothrix aspergillum* were described by Borowitzka and Vesk[50a] as having small vacuoles, little endoplasmic reticulum, few mitochondria, and several Golgi bodies with dilated cysternae. These are features characterictic of cells with high synthetic activity. Large meristematic cells up to 170 μm long occur in the branch apices of articulated plants like *Jania rubens,* whereas the smallest have been found in the perithallia of certain species of *Lithothamnium*.[76] Unusual meristematic cells that are free

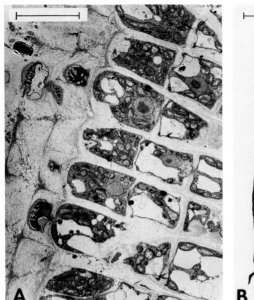 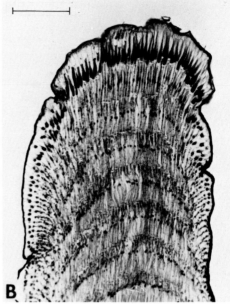

FIGURE 13. Primary meristems in articulated coralline branch apices. A. *Lithothrix aspergillum*. The meristematic cells are densely cytoplasmic whereas the older cells are more vacuolate. Some of the meristematic cells bear epithallial cells. At the bottom is the first division in pseudodichotomous branching. Scale = 10 μm. Courtesy M. A. Borowitzka and M. Vesk. B. *Bossiella cretacea*. The stain is hematoxylin. Scale = 100 μm. (Fig. 13B, Johansen, H. W., *Phycologia*, 15(2), 221, 1976. With permission.)

from one another occur in *Metagoniolithon* branch apices;[144,178] this is unlike the usual coherent arrangement in other entities.

Relative to position and function, there are two types of meristematic cells in coralline algae. Terminal meristematic cells are not surmounted by other cells. In aggregate these cells constitute a *primary meristem*.[465] This type of meristem occurs at crust margins, in which case they are marginal meristems, and at the apices of coralline branches, where they are called terminal meristems (Figure 13). At their distal poles meristematic cells are incompletely calcified and covered by a cuticle. Marginal meristems result in increasing areal coverage by crusts and terminal meristems result in branch elongation.

A second type of meristematic cell is intercalary and the tissue that several of them would comprise is an *intercalary meristem*. These meristematic cells are subapical in filaments where more distal epithallial cells are present. These intercalary meristematic cells are analogous to cambial cells in vascular plants. One of these cells may divide transversely to produce a new proximal vegetative cell and a distal cell that remains meristematic. Or, it may divide to produce a small distal epithallial cell and a larger proximal cell that remains meristematic. Intercalary meristems add to the thickness of crusts and intergenicula, and function at branch apices and crust margins in the Amphiroideae and Lithophylloideae.

Terminal meristems often display synchronous divisions. Thus a periodicity is exhibited that can easily be seen in sectioned tissues where adjacent cells are aligned in well-defined tiers (Figure 13). Environmental factors (e.g., temperature) may alter the rate of division as well as the lengths of the cells derived from the meristems.[235]

In branches of the Corallinoideae, two types of cells are derived from the terminal meristems, namely, intergenicular medullary cells and genicular cells. In *Corallina officinalis*, for example, after 14 to 18 tiers of intergenicular cells are produced, the

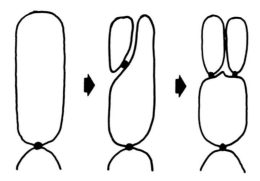

FIGURE 14. Diagram of pseudodichotomous branching in meristematic cell.[62]

meristem produces a tier destined to become a geniculum.[102] Following the formation of that single tier, the meristem reverts back to its usual role of producing intergenicular medullary cells. Precise coordination of the size of a terminal meristem and the formation of genicular cells occur as intergenicula grow and broaden prior to the production of branches (e.g., two in *Jania* and three in *Corallina*). When viewed from above, a terminal meristem will have widened considerably by the time the two or three branch buds arise, each subtended by a newly formed tier of genicular cells. In effect, a single meristem will have divided into two or three.

Intercalary meristems are apparently also affected by environment, as evidenced by what appear to be growth lines in perithallial tissue. Suneson[423] suggested that these lines reflect periods between growth spurts, analogous to annual rings in trees delimiting spring and summer wood. Furthermore, secondary tissues, including branches and conceptacles, are often produced from rejuvenating intercalary meristems, such as in cortices in *Bossiella orbigniana* and *Cheilosporum proliferum*.[235] In some thick crustose taxa the bulk of the tissue is apparently derived by regenerative growth of perithallial tissues.[76,81,262] This results in new, secondary hypothallia from which the plants are largely derived. This may be caused by growth on uneven substrates or injury, but details of how this rejuvenation occurs are unknown. It may be a standard part of development in *Lithophyllum* and *Clathromorphum*.

A normal part of development in the fronds of articulated coralline algae involves the transformation of intercalary meristems into terminal meristems when fronds are initiated. Cabioch[76] has shown that this involves the sloughing off of overlying epithallial cells in a localized area followed by growth of a frond bud by rapid divisions of the meristematic cells without the formation of additional epithallial cells.

The production of new meristematic cells has been observed by Cabioch,[62] who examined hypothallia of the bases of *Corallina officinalis* growing on microscope slides. When viewed from below, the marginal meristematic cells branch by first cutting off a new terminal cell by an oblique wall and then cutting off another terminal cell by a more or less transverse wall (Figure 14). The result is the production, by one meristematic cell, of two new meristematic cells of about equal size. This process is called *pseudodichotomous branching*. In this way the hypothallial cells of bases in *C. officinalis*, and probably crusts in many other species, branch in a marginal meristem of ever-increasing circumference.

EPITHALLIA

In most coralline algae epithallia are unistratose, although in some species they are

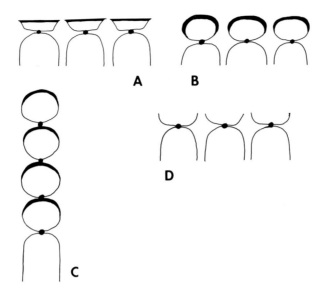

FIGURE 15. Diagrams of epithallial cells. A. The "eared" type in *Lithothamnium*. B. The rounded type in many genera, for example *Phymatolithon*. C. The type made up of several strata of cells in *Clathromorphum*. D. Dead cells in which the outer, uncalcified walls have been grazed away or otherwise lost.

multistratose when two or more epithallical cells terminate each perithallial filament. In *Clathromorphum, Lithophyllum orbiculatum*[7] and *L. incrustans*[185] epithallia are consistently multistratose.

Epithallial cells are short and have intensely staining, uncalcified walls at the thallus surface (Figure 15). Hence, the outer, uncalcified cell parts may be removed by grazing animals (e.g., limpets or chitons) leaving behind the calcified lower parts of the cells. When viewed by scanning electron microscopy,[34,181] these cell bases look like a pavement of cup-like depressions, each with a knob-like primary pit plug at its bottom (Figure 15). Usually, epithallial cells are intact and covered by a laminated cuticle which is often merged with the uncalcified cell walls.

In most species epithallial cells form over perithallial and cortical tissues, but are notably absent at the growing apices of hypothallial filaments at crust margins and medullary filaments at branch tips. As perithallial and cortical filaments are produced behind the margins and branch apices, epithallial cells are cut off by the meristematic cells (Figure 15).

The first coralline cell types to be studied with the electron microscope were lightly calcified surface cells making up epithallia and meristems. Giraud and Cabioch[185,186] and Borowitzka and Vesk[50] have revealed some unique features of epithallial cells, although their interpretations are not identical.

Giraud and Cabioch[185] described two types of epithallial cells. In one type, described from *Mesophyllum lichenoides, Corallina officinalis,* and *Jania rubens,* the cells form a unistratose layer covered by a more or less continuous cuticle. In the other type, studied primarily in *Lithophyllum incrustans,* there are two to four epithallial cells terminating each perithallial filament, a continuous cuticle is lacking, and the cells of different filaments are separate from one another.

Giraud and Cabioch[185] described epithallial cells as extremely active in secreting cuticular material by Golgi-produced vesicles discharging fibrillar material into the overlying cuticle. This seemingly logical conclusion was refuted by Borowitzka and Vesk[50]

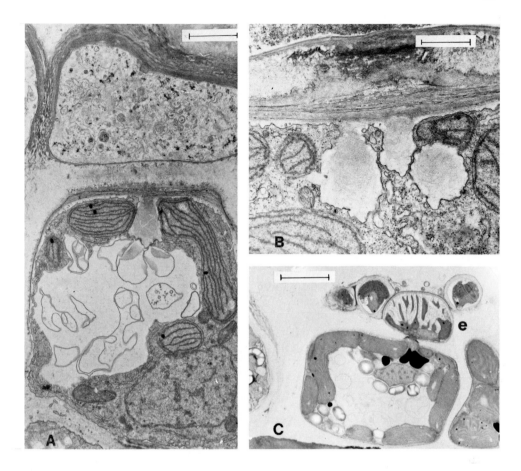

FIGURE 16. Electron micrographs of epithallial cells. A. Section through epithallial cell newly cut off by underlying meristematic cell in *Haliptilon cuvieri*. An uncalcified organic cell wall is being produced on outer surface of epithallial cell. Scale = 4 μm. Courtesy M. A. Borowitzka and M. Vesk. B. Cell wall ingrowths developing following death of overlying epithallial cell in *Haliptilon cuvieri*. Scale = 2 μm. Courtesy M. A. Borowitzka and M. Vesk. C. Epithallial cell (e) of *Fosliella* containing extensive cell wall ingrowths. Note the plastids in the underlying hypothallial cell. The coralline was growing on the angiosperm *Posidonia*. The small cells attached to the epithallial cell are epiphytic algae. Scale = 10 μm. Courtesy M. A. Borowitzka and M. Vesk.

who revealed that epithallial cells in *Corallina officinalis* and *C. cuvieri* (= *Haliptilon cuvieri*) have cell wall ingrowths extending into the cell lumina (Figure 16) and suggested that the previously reported vesicles were, in fact, ingrowths. This led Borowitzka and Vesk to hypothesize that epithallial cells are adapted to function as transfer cells with the ingrowths increasing the plasmalemma to cell wall contact area. They point out that analogous specialized epidermal transfer cells occur in a number of aquatic higher plants, but have not before been reported in algae. Furthermore, such modified cells also occur in other corallinaceous genera such as *Jania, Bossiella, Amphiroa*, and *Metagoniolithon*. Hence, it appears that epithallial cells are highly specialized to secrete cuticular material as well as transport nutrients from the surrounding seawater into the internal cells.

In *Corallina* numerous plastids occur in the distal parts of epithallial cells[50] and phycobilisomes have been reported in *Lithophyllum*.[185] Some mitochondria are also present, but the absence of starch grains led Borowitzka and Vesk[50] to speculate that much of the photosynthate is utilized locally or translocated to more internal cells where it is converted into starch.

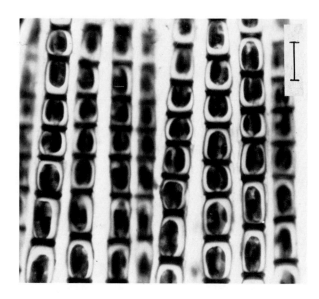

FIGURE 17. Vertical section of *Clathromorphum circumscriptum* showing epithallial filaments with dark-staining end walls (arrow). Scale = 10 μm. (Lebednik, P. A., *Syesis*, 9, 59, 1977. With permission)

In multistratose epithallia, the cells age as they are pushed farther from perithallial meristematic cells that are cutting off new epithallial cells. The oldest cells (nearest the surface) are more vacuolate, have smaller pastids, and are generally more degenerate, at least in *Lithophyllum incrustans*.[185] It appears that epithallial cells are sloughed off and renewed at intervals.[7] Meristematic cells underlying destroyed epithallial cells in *Mesophyllum lichenoides* produced a surge of cell wall polysaccharides.[78,79] This newly secreted polysaccharide became the surficial walls of new epithallial cells subsequently cut off below the dead epithallial cells. Borowitzka and Vesk[50] reported that when an epithallial cell dies the subtending meristematic cell lays down a cell wall below the pit-connection with the outer cell and, hence, seals it off. The sloughing process is probably important in keeping the calcified surfaces free of foreign organisms.

In *Clathromorphum* several strata of epithallial cells are usually present (Figure 15, 17), and unlike the epithallia in other corallines, the cells are entirely uncalcified. In the massive plants of *C. nereostratum*, Lebednik[262] found epithallia to be 4 to 14 cells thick. He also described structurally unique epithallial cell walls in *C. parcum*, with conspicuous end walls and undulate lateral walls.

Adey[8] has utilized the structural differences of epithallial cells in helping to delimit genera in the Melobesioideae. In *Lithothamnium* and *Sporolithon*, as *Archaeolithothamnium*, the uncalcified surficial walls are "eared"; that is, the uncalcified surface walls are flat-topped and flare out, whereas in *Phymatolithon* and *Leptophytum* these cells are simply rounded at the surface (Figure 15). As described above the multistratose types are characteristic of *Clathromorphum*.

Most calcified surfaces of coralline algae are covered by a cuticle of varying thickness which stain intensely in most stains and, where thick, appears as a white bloom on untreated surfaces. The thickness varies from 2μm to as much as 100 μm where conceptacle primordia are present (see conceptacle development in Chapter 5).

In *Clathromorphum* the cuticle extends as much as 1 cm from the substrate around the margin and back over the crust.[262] It may become 40 μm thick and is responsible for the shiny or glossy border seen in most plants. However, in older parts of the

Clathromorphum plants, a cuticle is usually disrupted or absent, an unusual condition that may be related to the thick epithallia in this genus.

Laminated cuticles are particularly conspicuous over articulated branch apices and over conceptacle primordia, although they also cover other calcified tissues. Bailey and Bisalputra [34] showed that cuticles in some articulated species consist of 6 to 7 layers, each being 0.25 to 1.5 μm thick, containing fibrils of unknown composition.

In a scanning electron microscope survey of thallus surfaces of 16 species in 9 genera, Garbary[181] observed differences that appear to be taxonomically distinctive. Surprisingly, in three species the cuticle remained intact and epithallial cells seemed to be absent, although in the other species cuticles and the upper parts of the epithallial cells were lost during preparation of the tissues. The three species apparently lacking epithallia were *Corallina officinalis* and two species of *Jania* (from the British Isles). In all except these three species the surfaces consisted of "concavities" formed by the bases of the epithallial cells, each with a central bump representing the basal pit-connection. The absence of epithallial cells in three species needs to be reinvestigated. Perhaps some thallus surfaces lack epithallia and the cuticle directly covers underlying meristematic cells; this would be true at branch apices. Garbary[181] found several structural patterns in the surfaces he studied: round to irregular shaped concavities throughout the surface, concavities in rows separated by grooves, and unevenly calcified concavities.

TRICHOCYTES AND MEGACELLS

Enlarged unpigmented cells in *Fosliella farinosa* (as *Melobesia farinosa*) were first called heterocysts by Rosanoff [364] and hair cells were first described for *Neogoniolithon notarisii* (as *Lithophyllum isidosum*) by Solms-Laubach.[412] Since these reports, numerous other researchers have described these structures in coralline algae, but now, as suggested by Cabioch,[75] the large, nonhair-bearing cells shall be called megacells and the hair-bearing cells shall be called trichocytes.

Most papers in which trichocytes and megacells are described pertain to fully-formed structures, but Cabioch[75] described their development in several taxa. Trichocytes and megacells are related, and the latter only appear following the formation of trichocytes. However, trichocytes alone occur in some species, such as *Jania rubens*. A trichocyte may originate in hypothallial, perithallial, or cortical meristems when a meristematic cell divides obliquely or transversely into two cells.[75] The newly formed cell walls do not become calcified, and the entire trichocyte is contained within the wall of the original meristematic cell (Figure 18). The ultimate cell becomes modified by the formation of an uncalcified hair, which grows out from the surface of the thallus. The hair is not cut off from its base, wherein remains the single nucleus; hence the hair cell somewhat resembles a carpogonium. The hair itself contains a prominent vacuole, is unpigmented, and is surrounded by a cellulose wall. In most taxa, the hair withers shortly after formation.

Cabioch[75] recognized three types of trichocyte and megacell development:

1. A simple type, such as in *Fosliella*, where trichocytes persist unaltered even after the hairs have been shed (Figure 19),
2. A more complex type, such as in *Jania rubens* and *Metagoniolithon radiatum* (as *M. charoides*), where trichocytes wither, megacells do not form and tissues dedifferentiate so that all evidence of the previous presence of trichocytes is lost in matured tissues,

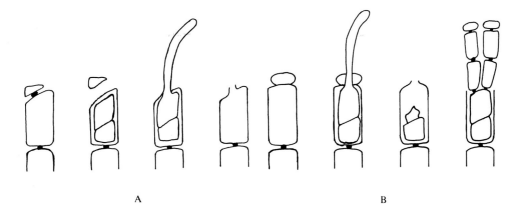

FIGURE 18. Diagram showing two types of trichocyte development. A. Note the cell division, loss of overlying epithallial cell, hair elongation and degeneration of hair and protoplast leaving an empty trichocyte wall. This type occurs in *Fosliella*. B. In this type the hair grows through the overlying epithallial cell, and cell degeneration and renewed meristematic activity obliterates evidence of trichocyte. This is the type present in *Jania rubens* and *Metagoniolithon radiatum*. Another type of development is like that shown here except that the overlying epithallial cells is sloughed off and renewed meristematic activity occurs above an ensuing megacell, which is retained as part of the vegetative tissue. This type occurs in *Neogoniolithon* and *Porolithon*.[75]

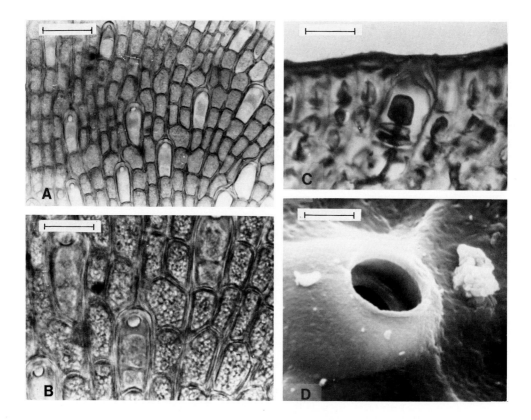

FIGURE 19. Trichocytes. A and B. *Fosliella farinosa* with terminal trichocytes. Note cell fusions in A. Courtesy Y. M. Chamberlain. A. Scale = 40 μm. B. Scale = 20 μm. C. Section through trichocyte in *Neogoniolithon*. Scale = 20 μm. (Fig. 19C, Johansen, H. W., *Phycologia*, 15(2), 221, 1976. With permission.) D. Scanning electron micrograph of raised pore representing remains of trichocyte in *Jania corniculata*. Scale = 4 μm. Courtesy D. J. Garbary.

3. The most complex type, such as in *Neogoniolithon notarisii, Porolithon onkodes* and *P. gardineri,* where trichocytes are short-lived, but where subsequent development results in megacells which persist for varying lengths of time in the coralline thalli.

Cabioch[75] described the production of *terminal trichocytes* in marginal meristematic cells of *Fosliella farinosa,* the continued growth of the thallus being accomplished as adjacent filaments grow around the trichocyte (Figure 19). Suneson[421] showed similar marginal cells in *F. limitata,* which can be considered as developing trichocytes. More common in *Fosliella* (e.g., in *F. limitata*) are *intercalary trichocytes* that result from the later transformation of a crust cell. Hair elongation is in an oblique direction and involves growth past the overlying epithallial cell.

Jania rubens (in Cabioch's type 2 group) was shown to possess trichocytes by Suneson[421] and later studied by Cabioch.[75] It is possible that environment plays a role in the extent to which trichocytes develop in this species as they were especially abundant in the summer in well-insolated localities. Development begins in the cortical meristem near branch apices with an oblique division of a meristematic cell into two cells. During this time the surmounting epithallial cell dies, and the hair produced grows directly through it, apparently dissolving the cell wall in the area of the pit-connection. Following elongation, the hair degenerates and the base of the trichocyte becomes meristematic, cutting off a new epithallial cell that replaces the penetrated one which sloughs off. In a scanning electron micrograph of an intergenicular surface of *Jania,* Garbary[181] showed a raised thickening that probably represented a trichocyte (Figure 19). As mentioned before, the older cortical tissues gave no indication that trichocytes had been produced.

In all essential details the development of trichocytes in *Metagoniolithon radiatum* (as *M. charoides*) is similar, except that here it was possible to see, below the meristem, open-ended remains of cell walls wherein trichocytes had been produced some time previously.[75]

Species in two crustose genera, *Neogoniolithon* and *Porolithon,* were also examined by Cabioch[75] for trichocyte and megacell development. In these genera, trichocytes are ephemeral and large megacells persist for varying lengths of time depending on the species. In *N. notarisii* two-celled and sometimes three-celled trichocyte ensembles form by transverse divisions and hair elongation (Figure 18). However, during development the cells making up the series enlarge, the lower cell becoming as high as 35 to 40 μm. In these species the large megacells are long-lived but, after sinking into the perithallus, they usually degenerate and endogenously originatin filaments of perithallial cells grow fo fill the void. In other species of *Neogoniolithon* variations occur, such as the development of vertical series of megacells as successive trichocytes are formed (Figures 19). Unlike *Porolithon,* there is considerable lack of unity in megacell formation in *Neogoniolithon,* and Cabioch[75] does not recommend placing much reliance on megacell arrangement as diagnostic of the genus.

Trichocytes are also ephemeral in *Porolithon,* being so drastically reduced and short-lived that they are hardly produced at all. But from these trichocytes ephemeral, as well as persistent megacells differentiate.[75] In both sexual and asexual plants of *P. onkodes* and *P. gardineri,* trichocytes originate in perithallial meristems in more or less circular uniform sori consisting of numerous enlarging cells, each of which eventually becomes two-celled. Here distinctions between the sexual and asexual plants appear. In sexual plants the trichocytes are ephemeral and disappear as new perithallial filaments grow from them. On the other hand, in asexual plants of *P. onkodes,* the two- to three-cell thick epithallus is sloughed off, reduced hairs develop, wither, and layers of mucilage overlie the trichocytes which by this time are two-celled because of

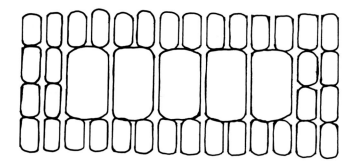

FIGURE 20. Diagram showing several megacells in *Porolithon* as they would appear in vertical section.

an oblique division occurring in each (Figures 18). However, these megacells developing from the trichocytes persist even as new perithallial cells and epithallial cells are produced from them, and they may easily be seen in older tissues that have been broken or sectioned (Figure 20). Ever since *Porolithon* was described,[164] it has been distinguished by lenses of megacells in the perithallia, but, as previously mentioned, the use of megacells in generic characterization has weaknesses.

Adey[3] reported rare trichocytes in *Phymatolithon rugulosum;* they may be an anomaly or a remnant of a vestigial ancestral trait. Lebednik[262] commonly found elongate, narrow, deeply staining cells embedded in thalli of *Clathromorphum reclinatum*, another melobesioid, and suggested that they represented hairs. Trichocytes have also been seen in *Dermatolithon*,[86-88,423] *Lithoporella* and *Mastophora*.[75]

The presence and structure of trichocytes and megacells have been used as diagnostic features in coralline algae for many years. For example, *Hydrolithon* has been characterized by "heterocysts present, arranged in vertical rows or scattered...,"[309] *Neogoniolithon* by "thalli with singly scattered heterocysts,"[303] and *Porolithon* by "...heterocysts grouped in short tranverse rows."[123] Many workers consider *Fosliella* to be distinguished from *Heteroderma* only by the presence of trichocytes in the former, and Mason[309] regarded the presence or absence of trichocytes "to be of fundamental phylogenetic significance." Others[86] doubt the genetic stability of trichocytes and megacells and suggest that their presence or absence may be environmentally induced.

SUMMARY

Although most vegetative cells in coralline algae are encased in calcium carbonate, there are some cell types which lack the mineral. Most coralline tissues are decalcified prior to study, a process which may alter cellular structure somewhat. Ultrastructural studies are being published at an increasingly rapid rate.

Calcified vegetative cell walls are three-layered. The thin layer adjoining the lumen is probably uncalcified. The bulk of the carbonate occurs in a thick middle layer. A calcified middle lamella joins contiguous cells. At least three species lack calcite. In the development of certain structures, such as genicula, decalcification occurs.

All vegetative cells are joined by primary pit-connections. These consist of the pores and pit plugs and, at least in two species, caps bracketing the plugs. In tissues where cell divisions are synchronous, the primary pit-connections form arching lines in sectioned material.

Secondary pit-connections develop between adjacent cells in the Lithophylloideae and Amphiroideae and rarely in other taxa. Except for their location on lateral walls, these structures resemble primary pit-connections. They are formed directly and involve formation of a small hole between the cells and, subsequently, the growth of a pit plug in the space.

Common in those taxa lacking secondary pit-connections are relatively large openings in the walls between adjacent cells. Nuclear control over cellular fusions has been suggested by Cabioch.[75] Several fused cells may form synetia with common protoplasts.

Except in certain reproductive cells, coralline nuclei are small and apparently similar to those in other red algae. Chromosome numbers of $n = 16$ and $n = 24$ have been determined for a few species.

Information on plastids, mitochondria, Golgi bodies, and starch grains is limited to a few recent reports of electron microscope studies. Staining bodies and crystals have been noted in light microscope studies.

In most species spore germination is of the *Dumontia* type and involves the growth of a small adherent disc. In a few specialized species, germination is of the *Naccaria* type where rhizoids grow out from the spore. Spores are large spherical cells that germinate quickly into four cells all in contact with the substrate. Some spores of *Haliptilon* and *Jania* may make only an ephemeral contact with a substrate after which the sporelings are swept into a host plant where they continue to grow.

As sporelings continue to grow past the four-cell stage, several patterns of cell division become evident. A basic distinction is present in the *Corallina* and *Amphiroa* types, and Figures 10 and 11 should be examined for these, as well as several other patterns that are less common. With further growth, small multicellular crusts with marginal meristems develop. The *Naccaria* type of germination occurs in a few epiphytic and parasitic species, such as *Choreonema, Amphiroa verruculosa,* and *A. rigida.*

Terminal meristems, in which the dividing cells terminate filaments, occur at crust margins and branch apices. Intercalary meristems are comprised of cells below filament apices, with the terminal cells constituting a skin-like epithallus over the meristem. In this type of meristem, the cells divide so as to add new cells both to the overlying epithallus as well as the underlying perithallus or cortex. Intercalary meristems result in increasing crust thickness and, in some taxa, in marginal growth and branch elongation. Divisions in some terminal meristems are synchronous and the primary pit-connections of derived cells are at the same level. Meristematic cells are dynamic, densely protoplasmic and, at least in terminal meristems, play a role in carbonate deposition.

Epithallial layers are single in most taxa and multiple in some, such as in *Clathromorphum*. The layers may be lost or removed by animals, in which case the underlying meristem cuts off new cells. Epithallial cell walls are usuallly uncalcified where the wall abuts the sea water and calcified on the side adjacent to the subterminal cell in the filament. Electron microscopy has revealed unusual ingrowths extending into the cell lumens which Borowitzka and Vesk[50] suggested facilitated transfer of materials in and out of the cells. The structure of epithallial cells as seen in longisection has been used to aid in delimiting some genera, such as *Lithothamnium* and *Phymatolithon.*

For many years enlarged trichocytes and megacells have been reported in several taxa, although they are absent in many others. In *Fosliella* terminal trichocytes originate in marginal cells and intercalary trichocytes in cells behind the growing margin. Development involves the growth of a hair from a single enlarged cell or from a cell complex. Hairs themselves may persist or may be shed shortly after growth, the enlarged cells may disappear by the ingrowth of new tissues or they may become buried and a permanent part of the thallus. Megacell grouping and position characterize certain genera such as *Porolithon,* but variation in megacell form also exists, making their taxonomic usefulness uncertain.

Chapter 3

STRUCTURE OF NONARTICULATED CORALLINE ALGAE

INTRODUCTION

In this chapter and the next, the development of young plants into mature thalli will be traced with an emphasis on how the plant types are organized. Possible evolutionary relationships based on vegetative structure are described when relevant, both within the subfamilies as well as among them. Growth and development are a function of the behavior of meristems and the differentiation of cells cut off from the meristems.

Tremendous anatomical diversity exists throughout the nonarticulated Corallinaceae, while at the same time the hardness conferred by the copious calcite in the cell walls and certain other structural features serve to unify the group. The diversity is also well demonstrated in the morphology of these plants (Table 1, Chapter 1). Environmental influences are being recognized with ever greater frequency in nonarticulated crusts[8] and in marl.[51,64] Evolution of structure, both in gross form and in internal organization, must, by its nature, be in the realm of calculated speculation; attempts to do this have been made by many students of extant and extinct coralline algae.

One reason that emphasis is placed on vegetative structures in coralline algae is that they exhibit a greater diversity of qualitative and quantifiable features than in most red algae. Individual filaments can often be traced for hundreds of micrometers if the tissue has been sectioned and stained properly. The fact that vegetative cells are approximately the same diameter in all parts of a tissue is helpful in this regard. Cell divisions are often synchronous after a group of meristematic cells have all grown to the same length. Dimensions of vegetative as well as reproductive cells are easy to obtain. Another reason for the advances made in studying coralline anatomy is that the plants are calcified and can be brought out of the sea and dried without distortion. Microtomed and stained sections of dried plants provide very adequate tissue preparations. Of course, delicate epiphytic crusts may shrivel with their hosts, and articulated coralline algae tend to fragment, but in the latter, intergenicular form is retained in dried plants. Hence, thousands of collections are housed in herbaria and many taxa have been named, although most older descriptions are based on external form only. This is one reason coralline taxonomy commonly involves the laborious unscrambling of 18th and 19th century nomenclature.

All coralline algae are filamentous and, in almost all instances, the filaments are aggregated into multiaxial thalli. Exceptions are the isolated filaments in sporelings of *Amphiroa rigida* and *A. verruculosa*[64] reduced parasitic forms such as *Choreonema*,[421] *Schmitziella*,[424] *Kvaleya*,[28] *Ezo*,[25] and the thin epiphyte *Fosliella farinosa* f. *solmsiana*.[76] In extant coralline algae the vegetative cells are 5 to 15 μm broad and extremely varied in length.

As cells in marginal meristems of sporelings begin to divide, new cells are added to filaments which grow approximately parallel to the substrate. These lowermost layers of filaments constitutue a hypothallus. Hypothallia may be monostromatic or multistromatic. In the latter type of hypothallus, the cells are produced by a marginal meristem in which cell divisions are irregular and seemingly random, or the divisions are synchronous so that the transverse walls of cells in adjacent filaments form arching lines when viewed in radial sections (Figure 1). This is called a *coaxial hypothallus*. In *Mesophyllum* this tissue forms a dominant part of the crust. Multistromatic hypothallia may be noncoaxial and relatively thin, such as in *Phymatolithon* (Figure 1).

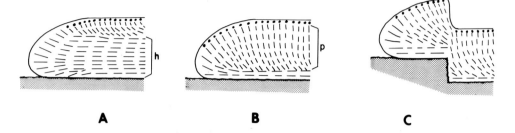

FIGURE 1. Diagrams of sections through crust margins showing organization of tissues. A. Coaxial hypothallus (h). B. Crust in which most of the tissue consists of perithallial filaments (p) arching up from a thin hypothallus. C. Secondary hypothallus derived from perithallus when irregularity in substratum (shaded) was encountered. In all figures epithallial cells are indicated by dots.

Hypothallial tissue may be produced from a marginal meristem derived originally from a sporeling; this is a *primary hypothallus*. Hypothallial tissue may also be generated from existing perithallial tissues or from primary hypothallia following disruption of this tissue.[262] This is called a *secondary hypothallus* (Figure 1). Lebednik[262] recognized that in plants of the massive *Clathromorphum nereostratum* most of the hypothallus may be secondarily derived. This may also be true in *C. circumscriptum* and *C. compactum*.[81] Cabioch[76] used the tendency for species to produce secondary hypothallia for taxonomic purposes, recognizing *Lithophyllum* a being disposed toward producing this tissue whereas *Pseudolithophyllum* is not.

Although some thin coralline crusts are made up solely of hypothallia and epithallia (except where conceptacles are produced), most crusts also have perithallia made up of filaments initiated by dividing hypothallial cells (Figure 2). Subsequent growth of perithallial filaments is by the activity of an intercalary meristem which, as described earlier, is almost always located below an epithallus. Perithallial filaments may be recognized in sectioned material because they are more or less perpendicular to the hypothallial filaments. Perithallial cells are shorter than hypothallial cells and often contain plastids, at least near the surfaces of thalli. In some species the perithallia may be thin and form only a minor part of the crust, whereas in others this tissue is much thicker than the associated hypothallus. If cell divisions in the perithallial meristem are synchronous, the resulting cells will be tiered.

Although the usual perithallial location in nonarticulated coralline algae is above a hypothallus and below an epithallus, there are other patterns. In *Metamastophora* and *Mastophoropsis* and in protuberances arising from crustose forms, the hypothallus is medulla-like and may or may not be bracketed by cortex-like perithallial filaments. In a description of *Metamastophora flabellata* Woelkerling[465] defined hypothallus not in relation to its position vis-a-vis a substrate, but rather as tissue derived from a primary meristem. Likewise, the perithallus was tissue derived from an intercalary meristem. In crusts which are unattached to a substrate for part of their extent, the hypothallus may be bracketed above and below by perithallia. Here the dorsiventral nature of the thalli results in the development of ventral as well as dorsal perithallia, for example in *Mesophyllum* and some species of *Clathromorphum*, the ventral perithallus is usually thinner.

For some anatomical studies in which crusts are sectioned, it is important to know the orientation of the cut. A crust originates at the point of spore germination and, at least initially, the hypothallial filaments radiate from that point. Thus a radical cut would present a different anatomical picture from that of a tangential cut, e.g., if a coaxial hypothallus were present (Figure 1).

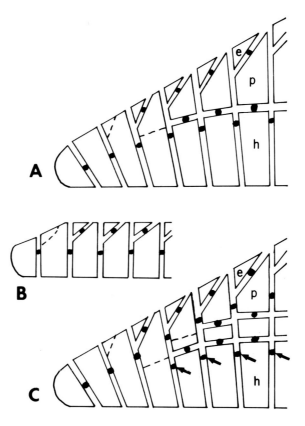

FIGURE 2. Diagrams of crust margins showing initiation of tissue types. A. Formation of hypothallus from which is derived small epithallial cells (e) and subsequently perithallial cells (p). Secondary pit-connections are not present. B. Perithallus lacking. C. The formation of epithallial and perithallial cells from the hypothallus. Here secondary pit-connections (arrows) are produced between the older hypothallial cells.

LITHOPHYLLOIDEAE

There is confusion over what genera to place in this subfamily and the distinctions among them, but, mostly on the basis of Cabioch's[76] studies of morphogenesis, *Lithophyllum, Tenarea, Ezo, Metamastophora, Dermatolithon,* and *Goniolithon* will be recognized. The last two genera are not recognized in some recent papers.[24,232] As discussed in the chapter on cytology, the Lithophylloideae are characterized by the possession of secondary pit-connections and one-pored tetrasporangial conceptacles.

It is possible to recognize two morphological series in this subfamily — one in which hypothallial and perithallial cells are as long as or only slightly longer than wide, including *Ezo* and *Lithophyllum,* and another with cells three or more times as long as wide, including *Dermatolithon, Tenarea* and *Goniolithon.* In each series Cabioch[76] has suggested, for certain species at least, that there are evolutionary gradations from species retaining primitive features (e.g., monostromatic hypothallia) to other species where the primitive are replaced by advanced characteristics as the plants grow to reproductive maturity. In the first series a part of *Lithophyllum* (which Cabioch calls *Pseudolithophyllum*) contains those species remaining primitive throughout development and in the second series *Dermatolithon* is reserved for the primitive species.

THE *LITHOPHYLLUM* SERIES

Lithophyllum orbiculatum, studied in detail by Adey[7] as *Pseudolithophyllum,* produces a monostromatic hypothallus from which perithallial filaments arise and, according to Cabioch,[76] this is the primitive state and the only anatomical condition in this species. A slightly more advanced condition is present in young plants in *Lithophyllum incrustans* where hypothallia are also unistratified for a time, a juvenile condition according to Cabioch.[67,76] Eventually, however, hypothallial growth ceases and perithallial cells at the margin of the crust produce a system of filaments growing more or less horizontally, much in the way of a multistratified hypothallus (Figure 1). As growth of this tissue continues cell divisions become synchronous so that the transverse cell walls become aligned and the crust has, for all intents and purposes, a coaxial hypothallus. Cabioch[76] considered this to be perithallial in nature and called it a "faux hypothalle". In fact, according to Cabioch[76] the genus *Lithophyllum* lacks true hypothallial tissue in older parts of the crust. Even in protuberances made up of distinguishable medulla and cortex, these tissues were considered perithallial. It would seem more reasonable, however, to consider the "faux hypotalle" as a secondary hypothallus, one that arises from perithallial tissue. Adey and Adey[20] did not find second hypothallia in most of the speciments of *L. incrustans* they examined and suggested that its formation may not represent normal development in this species. Hence, the importance of this phenomenon in separating *Pseudolithophyllum* and *Lithophyllum* remains an open question.

Until more evidence is available only *Lithophyllum* and the parasitic *Ezo* (treated later) will be considered here. *Lithophyllum orbiculatum* is a primitive species having only a primary hypothallus and *L. incrustans* is evolutionarily more advanced, having a prominent secondary hypothallus. Possibly the species of *Lithophyllum* are all either primitive or advanced, and the genus may be found to form two natural groups, in which case *Pseudolithophyllum* should be recognized.

Under certain conditions of damage to part of a crust, new tissues of *Lithophyllum* are regenerated and the juvenile *Pseudolithophyllum*-like crust will reappear for a limited time.[76] Sloughing of patches of perithallial tissues and the establishment of a new meristem below the former plant surface occurs in *Lithophyllum fasculatum.* Macroscopically the surface appears powdery when this occurs.

Ezo, with its single species *E. epiyessoense,* is parasitic on *Lithophyllum yessoense* and is closely related to its host.[25] The parasitic plants are white and reduced to small cushions of tissue bearing conceptacles. The hypothallia of both parasite and host are monostromatic and hence they belong among the primitive species of *Lithophyllum.*

THE *DERMATOLITHON* SERIES

In the second series in this subfamily, *Dermatolithon* is considered to be the simplest and most primitive genus.[76] Here the single-layered hypothallia consist of elongate cells usually oriented obliquely with respect to the substrate as viewed in radial section (Figure 2). In some species, hypothallial cells, each with a single, small, eccentrically positioned epithallial cell, consitute most of the the crust, at least the marginal parts (Figure 2). In other species, one or several tiers of elongated perithallial cells are also present. Marginal growth in this genus initially involves a regular cutting off of successive hypothallial cells by marginal meristematic cells. Shortly after each cell has been produced, a small epithallial cell is cut off at one side of the apex (Figure 2). At varying distances behind the margin, a perithallial cell is cut off from the hypothallial cell below the epithallial cell which by this time has enlarged slightly. Hence perithallial cells become intercalated between hypothallial and epithallial cells and, depending on

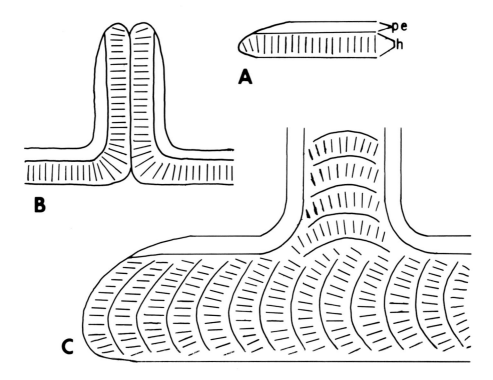

FIGURE 3. Diagrams of crust types in the Lithophylloideae. A. The simple type in *Dermatolithon* consists of a single layered hypothallus (h) and perithallial and epithallial tissues (pe). B. Two *Dermatolithon*-type crusts growing together so as to form an erect plate characterizing the genus *Tenarea*. C. Crust and protuberance of *Goniolithon* showing the extensive secondary hypothallus of elongate *Dermatolithon*-like cells.

the species, successive tiers of perithallia may be produced from the first formed (Figure 2).

Cabioch[76] pointed out the relationships of the dividing cells to the position of primary and secondary pit-connections in *Dermatolithon*. In one case the primary pit-connections between the hypothallial cells are low enough so as not to be lost when perithallial cells are produced from them (Figure 2). In another situation, these pit-connections are so high that they are left with the perithallial cells when they are produced and the only connections that occur between hypothallial cells are secondary pit-connections that form, as it were, to replace those lost to perithallial cells (Figure 2).

According to Cabioch,[76] species having a tendency for branches or erect plates that form when monostromatic *Dermatolithon*-like crusts grow together (Figure 3) exemplify the genus *Tenarea*, particulary as described for *T. undulosa* by Huvé.[215] This is considered an evolutionarily more advanced state. However, in several recent papers the distinctions between these genera that Cabioch recognized are not accepted, and some species of *Dermatolithon* have been transferred to the earlier-named *Tenarea*.[13]

Without apparent consideration of the distinctions given by Cabioch,[76] the genus *Goniolithon* has generally not been recognized in recent years.[232] However, she considered it to be related to *Dermatolithon* in the same way that *Lithophyllum* is related to her concept of *Pseudolithophyllum*. Thus, according to Cabioch, the mature crust consists entirely of perithallial tissue, a "faux hypothalle", as in *Lithophyllum*. However, this is a secondary hypothallus, hence, the thalli consist of both hypothallial and perithallial tissues. Huvé[216] showed that crusts of *Goniolithon papillosum* (as *Litho-*

phyllum) become relatively thick by the formation of secondary hypothallia made up of many layers of elongate *Dermatolithon*-like cells (Figure 3). In protuberances the secondary hypothallus constitutes a medulla in which the tiering is pronounced as in some articulated coralline branches. Under certain conditions of growth, regeneration of *Dermatolithon*-like juvenile crusts from well-formed crusts of *Goniolothon* does occur. Cabioch's[76] studies were on two Mediterranean species, *G. byssoides* and *G. papillosum*. She suggested that the thalli took on the *Goniolithon* aspect in warm waters where this genus occurs and remained *Dermatolithon* in cold waters. This problem is unresolved.

An unusual alga, *Tenarea (Dermatolithon?) tessellatum*, forming spiraled, knob-like protuberances on calcareous substrates below a depth of 10 m, occurs abundantly in the Hawaiian Islands.[283] The manner in which single marginal hypothallial cells divide results in a spiraled piling up of thalli; perithallia are lacking.

Woelkerling[466] has studied *Metamastophora flabellata,* the only species that he recognizes in the genus. Surprisingly, the entity has secondary pit-connections and a hypothallus of *Dermatolithon*-like cells. Therefore, this genus belongs in the Lithophylloideae rather than in the Mastophoroideae where it has usually been placed. The additional presence of cell fusions in *M. flabellata* indicates that the importances of using the types of cell connections in taxonomy at the subfamily level needs to be reevaluated.

MASTOPHOROIDEAE AND MELOBESIOIDEAE

These are two large subfamilies containing numerous genera of plants consisting of thin or thick crusts having lateral cell fusions rather than secondary pit-connections. (There are exceptions, however, such as a species of *Sporolithon* in which Cabioch[72] reported secondary pit-connections in addition to the usual fusions.) In all these general a commonness exists with respect to vegetative growth and development, even though the reproductive structures are so different that two subfamilies are recogniaed. According to Cabioch[76] thin crusts, such as in *Melobesia, Fosliella,* and *Heteroderma,* are more primitive than the more massive forms in *Lithothamnium, Phymatolithon, Clathromorphum,* and *Neogoniolithon.*

THIN CRUSTS

Many noncalcareous alga and marine angiosperms serve as hosts for thin (<200 μm) crusts. Such crusts also grow on rocks, but the difficulties in collecting and studying them has resulted in a paucity of work on them.[20] Some taxa may be exceedingly thin at the growing margins and the conceptacles may protrude conspicuously, whereas in other, slightly thicker forms, the conceptacles may be approximately flush with the thallus surface. Some thin crusts may be host specific or partly so, for example, *Fosliella lejolisii* on eel grass *Zostera marina* L. Others grow seemingly on any host, for example, *Heteroderma nicholsii.*

Fosliella, Heroderma, and *Melobesia* are thin crusts that in some species, such as *F. limitata, F. lejolisii,*[84] and *M. membranacea* lack perithallia in vegetative parts, As described by Cabioch[76] marginal growth here occurs when meristematic cells cut off hypothallial cells, each of which subsequently cuts off a small eccentrically positioned epithallial cell (Figure 4). In thicker species perithallial cells are cut off from the newly produced hypothallial cells, and these in turn cut off epithallial cells. As these crusts increase in diameter, marginal cells occasionally divide pseudodichotomously and thus increase the number of hypothallial filaments in the expanding margin.

A modification of the uniform crust described above occurs in *Fosliella farinosa* f.

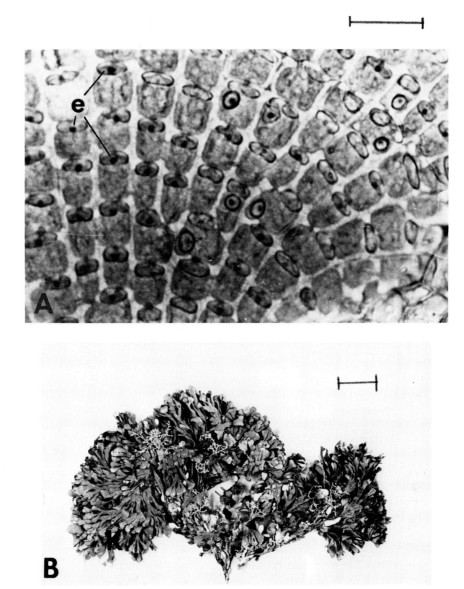

FIGURE 4. A. Surface view of *Fosliella* in which a perithallus is lacking. Note a small epithallial cell (e) overlies each hypothallial cell. Fusions are present between some of the hypothallial cells. Scale = 20 μm. Courtesy Y. M. Chamberlain. B. *Mastophoropsis canaliculata*, a ribbon-like thallus attached at one point (arrow). Scale = 1 cm. Courtesy W. J. Woelkerling.

solmisiana where the hypothallial filaments are not laterally contiguous throughout. There is no marginal meristem in this alga, and new hypothallial filaments are formed from initial cells that may be several cells behind a filament apex.[76] A new filament often grows at approximately right angles to the filament from which it arises. Filaments cohere when conceptacles are produced.

According to Cabioch,[76] there are some species structured like *Fosliella* but lacking epithallia. Species having this simplification of the standard *Fosliella* structure have been placed in the poorly known genus *Litholepis*. Cabioch[76] though that perhaps *Litholepis* should be considered a subgenus under *Fosliella*, and Adey[13] suggested that *Lith-*

olepis is closely related to *Lithoporella* and *Heteroderma*. Further study is needed.

In other species of *Fosliella* (e.g., *F. valida*) and *Heteroderma* (e.g., *H. zonalis*) perithallial filaments are formed. These crusts may become as much as 100 μm thick in parts lacking conceptacles.[116,117]

There have been some recent studies of *Lithoporella*,[273-275,303] a thin crustose entity in which a perithallus is lacking in most parts of the plant. Like *Litholepis, Lithoporella* crusts tend to overgrow one another, and hence they superficially appear thicker than they really are. Lemoine[275] mentioned plant consortiums consisting of 5 to 26 strata. Adey[13] pointed out that *Lithoporella* has cells larger than in most extant coralline algae; they are usually more than 10 μm in the smallest dimension. This contrasts with the small-celled *Litholepis*. Unusual features of *Lithoporella* are

1. The presence of unicellular rhizoids.[76,275,303]
2. A tendency for crust margins to grow downward towards lower crusts.[76,273-275]
3. The penetration of the tissues of lower crusts by cells of the marginal meristem of plants immediately above.[273-275]
4. A tendency for crusts to overgrow one another.

RIBBON CORALLINES

Another growth form for coralline algae is thin dichotomously branching crusts that are free from a substrate for most of their extent. Plants exhibiting this form are attached to hard surfaces by holdfasts at one end of somewhat erect, branched dorsi-ventral, ribbon-like thalli (Figure 4).

These algae probably evolved from crustose forms. During evolution of the erect habit the lower surfaces in *Mastophoropsis* and the stipes and older veins in *Metamastophora* became endowed with epithallia and perithallia (cortices). The ventral surfaces contain fewer plastids and are generally more reduced than the upper. Also, the usual restriction of conceptacles to the upper surface clearly reveals the dorsi-ventral nature of the thalli. Woelkerling[446,465] recently showed that such growth forms evolved not only in the Mastophoroideae (*Mastophora* and *Metamastophora*) but also in the Melobesioideae (*Mastophoropsis*) and Lithophylloideae (*Metamastophora*).

Primary branching is dichotomous or polychomotomous and in one plane. Adventitious branches may also arise from dorsal surfaces in *Mastophoropsis;* usually they are associated with damaged or severed axes.[446]

Mastophora appears to be closely related to *Lithoporella* in spite of the unique habits exhibited by these two genera. Lemoine[275] noted similarities in several important features, but described *Lithoporella* as strongly calcified and the other genera as appearing cartilaginous. In fact, ribbon corallines are more or less flexible and appear to have evolved a means of projecting their conceptacles higher into the water, as have the articulated coralline algae.

Most meristematic activity occurs at the branch tips where the axial filaments are produced. In *Mastophora*, marginal meristematic cells cut off hypothallial and epithallial cells and, as shown by Cabioch[76] for *M. macrocarpa*, perithallial cells only near conceptacles. Trichocytes are common in this species as are fusions between hypothallial cells. Structurally, *M. macrocarpa* is very much like *Lithoporella*. Suneson,[426] Cabioch,[76] and Woelkerling[465] have shown that in *Metamastophora flabellata* (the type of the genus) perithallial cells are formed even in vegetative parts of the crusts, the filaments becoming 1 to 4 or 5 cells long. A perithallial meristem does not develop (except where conceptacles form), however, and the perithallial cells are cut off from hypothallial cells.[76] The apices and margins of the plants lack perithallial cells and, anatomically, are similar to *Mastophora*. In the lower parts perithallial tissue builds up to form a midrib which grades into a short stipe.

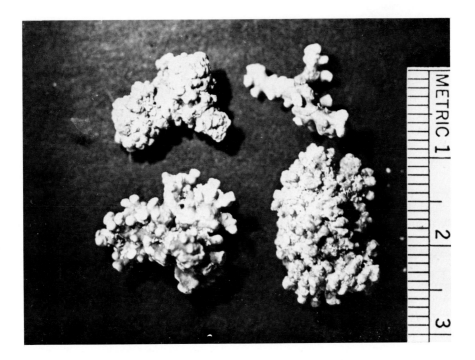

FIGURE 5. Samples of marl from Whalebone Bay, Bermuda. Scale in centimeters. Collected by M. Hopkins and E. J. Johansen.

THICK CRUSTS

There are several genera in which most species are epilithic and produce crusts that become more than 200 μm thick (Figure 5), with some massive forms such as *Clathromorphum nereostratum* becoming 10 cm or more thick.[262] The thickness of these crusts depends on the amounts of hypothallial and perithallial tissues present; either may form the bulk of the crust. The extent to which these genera have been studied differs markedly, with *Clathromorphum*[14,15,81,262] and certain species in *Phymatolithon*,[3,8] *Leptophytum*,[8] *Neogoniolithon*,[76,303] *Lithothamnium*,[8,11,20,76] and *Porolithon*,[8,76] recently and intensively studied. The generic name *Lithothamnium* has been used in earlier works to include taxa that are now recognized as belonging to other genera. Hence, only more recent papers dealing with *Lithothamnium* should be considered seriously, with Adey[8] having circumscribed the genus as it is presently known by most workers. Several genera in which the plants are thick crusts are less well known: *Antarcticophyllum*, *Mesophyllum*, and *Sporolithon*.

Neogoniolithon and *Porolithon* are two closely related genera with thick, many-layered hypothallia and fairly copious perithallia. Some species are relatively smooth crusts whereas in others variously expressed protuberances develop. Basically identical developmental features characterize several genera in the Melobesioideae: *Lithothamnium, Clathromorphum, Phymatolithon, Leptophytum, Mesophyllum*, and the newly described *Antarcticophyllum*. *Hydrolithon* (in the Mastophoroideae) is also considered here, even though the hypothallia are unistratose.

Marginal growth in all these genera is by divisions at the apices of hypothallial filaments. The lowermost filaments tend to grow parallel to or curve toward the substrate and the upper filaments curve upward, branch pseudodichotomously and become perithallial in nature.[76] In almost all species epithallial cells are cut off behind the crust margin as a perithallial meristem develops. Perithallial cell length varies considerably among species, such as in *Neogoniolithon* where they may be 25 to 30 μm in *N. notarisii* or only 5 μm in another, undetermined species.[76]

In many of the species of *Neogoniolithon* and *Porolithon* protuberances form. Like the origin of fronds in the Corallinioideae, protuberance initiation is in the perithallial meristem.[76] As a protuberance grows, the internal filaments become medulla-like and the peripheral filaments cortex-like.

The relative amounts of hypothallial and perithallial tissue in crusts varies from species to species. Examples given by Cabioch[76] are *P. onkodes,* where most of the crust is composed of perithallus and neither it nor the hypothallus is produced by synchronous divisions, and *P. gardineri,* where the perithallus is thin and the hypothallus thick and coaxial. The species of *Clathromorphum* may also be divided into those with thick hypothallia and those with thin hypothallia.[262]

Most crustose species grow in a manner so that they remain closely appressed to the substrate. Some species, however, have margins that are free of the substrate to varying extents. Notable for the characteristic of being loosely associated with a substrate are *Mesophyllum* and various species scattered among the other genera. The enormous thalli of *Clathromorphum nereostratum* produce margins that sometimes continue to grow unattached for many centimeters and morphologically resemble bracket fungi.[262] Lebednik[262] found that grooves occurring in the surfaces of thalli of *Clathromorphum* correspond to stained lines extending vertically through the crusts in sectioned material. The surface grooves result from the interaction of filaments that are either branching or fusing as the crusts respond in growth to the contours of the substrate. He felt that these grooves in *C. nereostratum* probably represent an adaptation allowing the crust to grow up or down without splitting or crushing the filaments.

UNATTACHED CORALLINE ALGAE

Several species of *Lithothamnium, Phymatoliton, Neogoniolithon,* and *Lithophyllum* occur unattached as marl or rhodoliths on ocean floors, sometimes in great numbers. In the structural context marl are irregularly shaped, coralline algae usually made up of several protuberances projecting from a common center (Figure 5). They are only one or a few centimeters in the longest dimension. Probably marl originates when fragments bearing protuberances are broken from attached crusts, rhodoliths, or other pieces of marl. The pieces continue to grow, the eventual form depending greatly upon the agitation they they undergo (see Chapter 9).

Rhodoliths are coralline nodules that are usually larger than marl and that have grown by layers around a center, or nucleus. They are roughly spherical although the surfaces are often roughened by protuberances. The distinction between marl and rhodoliths is often unclear, and it is sometimes necessary to section a rhodolith in order to verify its nature. Efforts need to be made to devise clearer definitions of unattached coralline algae.

EPIPHYTIC CORALLINE ALGAE

Several genera contain plants that are visibly modified for living on plants, including other coralline algae (Table 1). These modifications range from reduced vegetative systems and loss of pigments in plants that produce scarcely more than conceptacles, to button-like forms that are little different from their epilithic relatives. These modified plants have evolved to fill their particular niches over millions of years in various phylogenetic lines, hence, today, they belong to different subfamilies in the Corallinaceae. Other species contain no visible modifications, but nevertheless apparently grow only on other plants.

Table 1
NONARTICULATED CORALLINE ALGAE THAT APPARENTLY GROW OBLIGATELY ON OTHER CORALLINE ALGAE

Taxa	Hosts	Relationships	Ref.
Ezo epiyessoense	Lithophyllum yessoense	Parasitic, unpigmented	25
Kvaleya epilaeve	Leptophytum laeve	Parasitic, unpigmented	28
Choreonema thuretii	Jania, Haliptilon, Cheilosporum	Parasitic, pigmented conceptacles	421
Chaetolithon deformans	Jania	Parasitic	168, 253
Clathromorphum parcum	Calliarthron tuberculosum	Epiphytic	23, 262
Clathromorphum reclinatum	Calliarthron, Bossiella, Corallina^a	Epiphytic	23, 262
Mesophyllum conchatum	Calliarthron, Bossiella, Corallina	Epiphytic	23
Dermatolithon corallinae	Corallina	Epiphytic	423

^a Also reported on species of *Ahnfeltia*, *Laurencia*, and *Prionitis*.[262]

Four species are modified for an epiphytic existence. Two species of *Clathromorphum* and one of *Mesophyllum* produce irregularly-shaped crusts on host plants in the eastern Pacific Ocean (Figures 6, 7). *Synarthrophyton patena* is a similarly constructed entity in the southern hemisphere (type locality in New Zealand). In these epiphytes margins are free and the plants have a distinctive mushroom-like appearance which induced Mason[309] to erect the genus *Polyporolithon* for these four species. *Synarthrophyton patena* grows on *Ballia callitrica* (C.Agardh), Kützing, and other noncalcareous algae whereas the other three taxa usually occur on articulated coralline algae.

Clathromorphum parcum, for many years known as *Lithothamnium parcum* and *Polyporolithon parcum*, is a mushroom-shaped plant that grows only on the articulated *Calliarthron tuberculosum* from Northern Washington to Central California.[1] The smooth-topped crusts, up to 22 mm in diameter and 3 mm thick, are attached to the host by short, approximately central stalks, the bases of which are deeply embedded in cortical tissue of the host (Figure 6). Adey and Johansen[23] have shown that when spores of *C. parcum* germinate near the apices of *C. tuberculosum*, the initial growth of sporelings is upward rather than horizontal as in most coralline algae. The upward growth is matched by cortical growth of *C. tuberculosum* until eventually, the epiphyte has a firmly embedded base or "foot" (Figure 6). Continued growth is lateral and the conceptacle-containing, flat-topped structure is formed. Progressive stages in the co-development of epiphyte and host are shown in Adey and Johansen.[23]

Detailed studies of *C. parcum* by Lebednik[262] show that this species differs from other species of *Clathromorphum* in two important respects: (1) Epithallial tissue surrounds the thallus margin as well as the ventral and dorsal surfaces, hence, the entire crust that is exposed to seawater is covered by an epithallus. The ventral and dorsal epithallia are 3 to 5 cells thick, but along the margins there are usually only 2 to 3 epithallial layers. Ventral cells have fewer plastids than do dorsal cells. Lebednik[262] speculated that the marginal epithallus is an evolutionary adaptation to relatively high light intensity, lack of competition afforded by its epiphytic habit, or its particularly

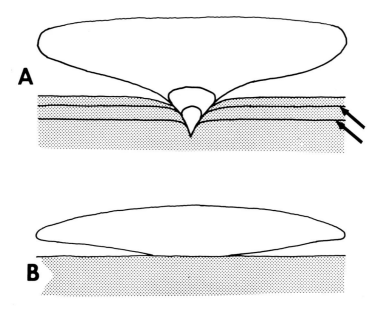

FIGURE 6. Diagrams of sections of nonarticulated coralline algae growing on articulated coralline algae (shaded). A. *Clathromorphum parcum*, showing embedded foot. Successive growth of epiphyte and host indicated by lines (arrows). B. *Mesophyllum conchatum*. The plant is attached to the host by a pad of tissue.

unique early development. (2) According to Lebednik's[262] interpretation, growth of the primary hypothallus is suppressed and the bulk of the plant consists of perithallial cells (Figure 7). However, epifaunal irritation on the ventral surface can result in the production of a secondary hypothallus from the perithallial meristem.

Two other species, *Clathromorphum reclinatum* and *Mesophyllum conchatum*, also grow on articulated coralline algae, but no foot is produced and lateral hypothallial growth commences immediately on germination.[23] *Clathromrphum reclinatum* usually grows on *Bossiella* and *Corallina*, although it has also been found on *Ahnfeltia plicata* (Hudson) Fries. Unlike the relatively thick *Clathromorphum parcum* (up to 3mm), these plants are thin (less than 1 mm) and tend to be more undulate and irregular with *C. reclinatum* sometimes becoming wrapped about the host branches.

C. reclinatum is less modified for its epiphytic condition than is *C. parcum*. As pointed out by Lebednik,[262] the bulk of the plant is made up of a thick primary hypothallus. Although the maximum crust diameter is 1.5 cm, the primary hypothallus is produced by a marginal meristem that is in all essentials like that in other species of *Clathromorphum*. The copious hypothallus consists of ascending filaments that become perithallial and descending filaments that end in thick walled, club-shaped cells at the ventral thallus surface. Once a perithallial meristem is established behind the plant margin, a thin epithallus 1 to 2 cell layers thick is produced. This epithallus is thinner than others in the genus *Clathromorphum*. Lebednik[262] suggested that the relatively thick cuticle in this species inhibited the growth of a more copius epithallus. The perithallial meristem is unusual in that it consists of large cells amply endowed with plastids and starch grains and, in fact, constitutes the primary photosynthetic tissue of the plant. This was one of the features which induced Adey and Johansen[23] to erect the genus *Neopolyporolithon* for this species.

Early development has not been studied. However, the crust is attached to its host by a simple pad of tissue. Additional adherent pads are formed on occasion by second-

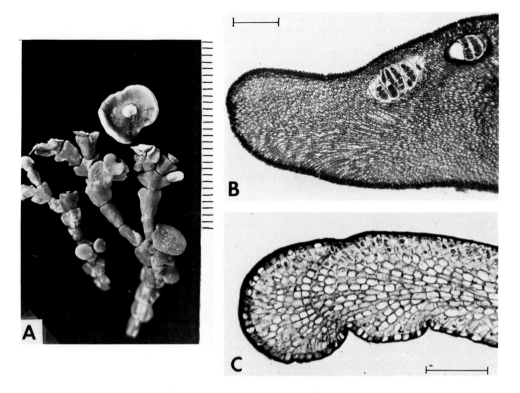

FIGURE 7. Epiphytic coralline algae. A. *Clathromorphum parcum* on *Calliarthron tuberculosum*. Note that the upper specimen shows a ventral surface and central stalk. Scale in millimeters. B. Vertical section through *Clathromorphum parcum*. The epithallus is densely stained and covers both upper and lower surfaces and the margin. Scale = 250 μm. C. Margin of *Synarthrophyton patena*. The epithallus is single-layered. Scale = 100 μm. (Johansen, H. W., *Phycologia*, 15(2), 221, 1976. With permission.)

ary hypothallia. Lebednik[262] found no evidence of the penetration of host tissue such as had been reported by Masaki and Tokida.[306]

The main features of *Mesophyllum conchatum* were described by Mason[309] (as *Polyporolithon conchatum*) and Adey and Johansen.[23] Like *Clathromorphum reclinatum*, and unlike *C. parcum*, *M. conchatum* is attached to its host by a pad of hypothallial tissue and the primary hypothallus, which is coaxial as in other members of this genus, forms the major part of the plant (Figure 6). The thalli are circular to semicircular and as large as 3 cm in diameter, but only attain a thickness of 1 mm.[1] The margins lack epithallial cells, these first appearing on the upper surface behind the margin of the crust.

Synarthrophyton patena, although first described in 1847[199] as *Melobesia patena*, has subsequently been placed in *Mastophora, Lithophyllum, Lithothamnium, Polyporolithon, and Mesophyllum* and has recently been studied by Townsend.[435a] The plants are usually plate-like, although they sometimes enfold host branches. They are up to 2 cm in diameter, but they rarely become more than 250 μm thick. When the plants grow on filamentous hosts such as *Ballia* or *Cladostephus* (phaeophyta), they become attached by hypothallial extensions which intersperse with host tissue. They sometimes grow on articulated coralline algae, in which case they form attachment pads.

The thalli of *S. patena* are made up of thick coaxial hypothallia and relatively thin perithallia. The coaxial hypothallia and rounded epithallial cells suggest *Mesophyllum*, but because the plants also posses certain *Lithothamnium*-like reproductive characters,

Townsend[435a] recognized the new genus. As pointed out in a previous chapter, this taxon is unusual in having secondary pit-connections as well as cell fusions.

Adey and Johansen[23] noted that the epiphytic species in *Clathromorphum* and *Mesophyllum* had larger cells than epilithic species with which they could be compared. Cells of *M. conchatum* are considerably larger than those in *M. lichenoides,* a species which is usually epilithic. Furthermore, although the data are not ample, Adey and Johansen[23] noted that two epiphytic plants of *M. lichenoides* had larger cells than did epilithic plants in the same species.

In these epiphytic species there appears to be no transfer of material between host and epiphyte and probably the co-evolution that resulted in these relationships proceeded for reasons other than nutrition. Furthermore, they are as pink as epilithic coralline algae. They were termed "hemiparasitic" when they were included in the genus *Polyporolithon.*[309] Another species of *Mesophyllum, M. lamellatum,* in addition to being epilithic, grows on articulated coralline algae, but they form extensive crusts that overgrow numerous fronds at once.[1]

PARASITIC CORALLINE ALGAE

Five corallinaceous genera contain plants that are greatly modified for an existence that is dependent on specific hosts that serve as substrates. These genera are: *Kvaleya, Ezo, Choreonema, Chaetolithon,* and *Schmitziella.* The first three grow only on corallinaceous hosts, and the last is restricted to the chlorophtye *Cladophora.* The Australian melobesioid *Chaetolithon,* with only a single species, *C. deformans,* is represented by small plants growing on *Jania* sp. and causing the apices to be deformed. In fact, according to Fritsch,[168] infected intergenicula become so proliferous that only the conceptacular pores of the parasite remain uncovered. However, so little is known about *Chaetolithon,* that it shall not be treated further.

The striking parasite *Kvaleya epilaeve* was described from the North Atlantic (type locality is Norway; also found in Iceland and colder parts in the western North Atlantic) by Adey and Sperapani.[28] A member of the Melobesioideae, it grows only on the crustose *Leptophytum laeve,* also a member of this subfamily. In fact, in spite of its greatly reduced size, the parasite is similar to its host in taxonomically important features and thus qualifies as a good example of a rhodophycean adelophoparasite.

Plants of *K. epilaeve* consist of clusters of conceptacles with small vegetative lobes (Figure 8).[28] They are unpigmented. A hypothallus, several cell layers thick, and a perithallus are present, but epithallial cells are mostly lacking. As young plants grow over the host by a marginal meristem, some of the apical cells become transformed into unicellular haustoria. These haustroia may penetrate as much as 25 μm into host tissue, and they sometimes penetrate host cells. Even though numerous plants of *K. epilaeve* may occur on a single crust, there is apparently no harm caused.

Ezo (named for Ezo, the old name for Hokkaido), with its single species *E. epiyessoense,* is parasitic on the closely related *Lithophyllum yessoense,* a common species in shallow subtidal areas in northern Japan.[25] Superficially similar to *Kvaleya, Ezo* also consists of small cushions of white tissue, bearing conceptacles, and producing host-penetrating haustoria. Like its host, *Ezo* has a single-layered hypothallus, abundant secondary pit-connections, and single-pored asexual conceptacles with prominent columella. Unicellar haustoria arise occasionally from hypothallial meristematic cells and may extend between host epithallial cells (there are 3 to 5 epithallial cell layers in *L. yessoense*) and even into the perithallial meristem. Here haustoria and meristematic cells may come into close contact or they may form "complete fusions".[25]

Evolution has thus produced the similar adelphoparasites *Evaleya* and *Ezo* in different subfamilies in widely separated parts of the world. A third parasite, *Choreonema,*

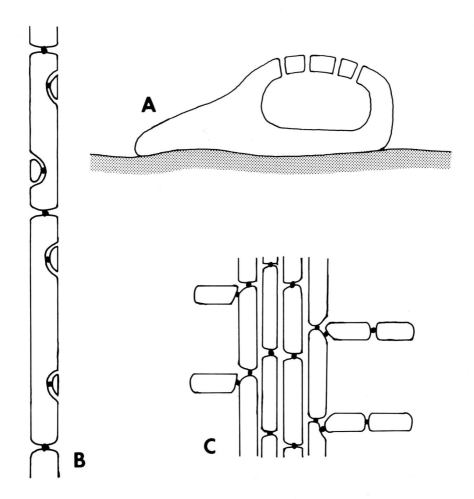

FIGURE 8. Diagrams of parasitic coralline algae. A. *Kvaleya epilaeve*. The section passes through a multipored conceptacle and a vegetative lobe. Host tissue is shaded. Adapted from Adey and Sperapani.[28] B. A filament of *Choreonema thuretii* as it appears growing among medullary filaments of a host. The small cells are interpreted as vestigial epithallial cells. Adapted from Suneson.[424] C. *Schmitziella endophloea*, showing the central filaments and the peripheral filaments arising from them. Adapted from Suneson.[421]

is also the result of a long evolutionary history, but it differs in certain respects from the first two.

Choreonema is represented by a single species, *C. thuretii*, which is present in widespread areas such as Atlantic Europe, the Mediterranean Sea,[56] the western Indian Ocean (pers. obs.), and the eastern North Pacific Ocean.[1] It is an example of long adaptation to an existence restricted to living with branches of *Jania, Haliptilon,* and *Cheilosporum*. Its probable restriction to hosts in these three genera has led to the suggestion that *C. thuretii* represents a rhodophycean adelphoparasite derived from members of the tribe Janieae.[236] Bressan[56] however, reported that it also occurs on *Corallina officinalis* in the Mediterranean Sea. Thus, until more is known about its anatomy and host relationships, it should be retained in the Mastophoroideae.

The vegetative portions of *C. thuretii* are made of filaments of long, unpigmented cells (= hypothallial cells) that grow individually among the medullary filaments of host intergenicula.[421] Small cells may be budded from these vegetative cells (Figure 8); they have been interpreted as vestigial epithallial cells.[421] Grouping of cells at the host surface occurs prior to conceptacle formation.

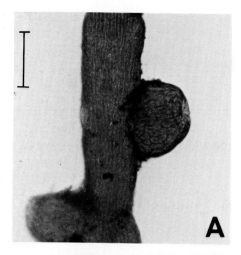

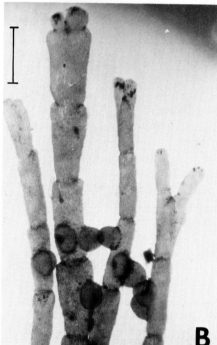

FIGURE 9. A and B. *Choreonema thuretii* on *Haliptilon squamatum*. Only the conceptacles of the parasite are externally visible. A. Scale = 250 μm. (Fig. 9A, Johansen, H. W., *Phycologia*, 15(2), 221, 1976. With permission.) B. Scale = 500 μm.

Nothing is known about the manner of infection except for the studies of spore germination by Cabioch[76] described earlier. The host plants are affected, with branch endings in infected *Haliptilon squamatum* in western Ireland contorted and dwarfed (Figure 9). The plants are parasitic, except that the small conceptacles which come to protrude from intergenicular surfaces of the host are pigmented (Figure 9). In fact, based on the appearance of *Choreonema* conceptacles that were not recognized as belonging to a parasitic organism, *Jania verrucosa* was described as new.[256]

Perhaps the most aberrant coralline algae are the two endophytic species in *Schmitziella*, *S. endophloea*, restricted to *Cladophora pellucida* Kützing in Atlantic France, the British Isles and occasionally in the Mediterranean Sea and *S. cladophorae*, restricted to *Cladophora feredayi* in New Zealand. These species are among the few coralline algae that do not calcify. Suneson[424] found that *S. endophloea* from Great Britain consists of filaments in single layers growing between the outer cell wall layers of the host. The thalli consist of two types of filaments (Figure 8): (1) primary filaments of elongate cells sometimes growing in contact with one another and (2) secondary filaments of short, irregular shaped cells which arise from the primary filaments. No epithallial cells are present.

The way in which *Cladophora* becomes infected is not known. Batters[36] suggested that entrance might be through pores in the host walls made prior to zoospore discharge. Specimens of *Cladophora* infected with *S. endophloea* are often also infected with epiphytic crustose species such as *Melobesia membranacea*.[424] Often the epiphyte occurs lower on the host branches and the endophyte higher, but sometimes *S. endophloea* grows underneath the *Melobesia*. Infected branches are rose-pink in color.

SUMMARY

Much structural diversity exists among those coralline algae lacking genicula, but in most entities a hypothallus, perithallus, and epithallus may be recognized. Hypothallia are primary and derived directly from spore-generated crusts, or secondary and derived from other tissues, such as perithallia. Hypothallia are thin or thick and, if produced by synchronous divisions, coaxial. Perithallia may be thick, thin, or absent except around conceptacles.

In the Lithophylloideae, two series, each with supposedly primitive and advanced genera, may be recognized. Morphogenesis may be studied by analyzing the production of the precisely organized tissues with reference to the positions of primary and secondary pit-connections.

Lithophyllum contains primitive species with monostromatic hypothallia and advanced species with thick secondary hypothallia.

Dermatolithon is a simple, primitive genus that perhaps led to the more advanced *Tenarea* and *Goniolithon*. In *Tenarea* plates form by the union of simple *Dermatolithon*-like crusts and in *Goniolithon* the massive crusts are tiered and produce protuberances.

The Mastophoroideae and Melobesioideae lack secondary pit-connections and exhibit a variety of structural types. Thin forms are possibly more primitive than thick forms.

Thin crustose species in *Fosliella*, *Heteroderma*, and *Melobesia* may lack perithallia in vegetative parts and are often epiphytic. In *F. farinosa* f. *solmsiana* the hypothallial filaments are not contiguous. *Litholepis* and *Lithoporella* have unique overgrowing habits.

Some thin forms of nonarticulated coralline algae have evolved erect, branched, ribbon-like thalli that are often dorsiventral. These occur in two subfamilies, the Mastophoroideae, Melobesioideae, and Lithophylloideae.

Several crustose genera have thick hypothallia or perithallia (or both), and even within a single genus the species may vary in this regard.

Marl are free-living, branched, nodular forms which may originate when protuberances break away from attached crusts and continue living. Roughly spherical rhodoliths form when crustose coralline algae grow around nuclei and become multilayered after many years of agitation and growth.

Noncorallinaceous marine plants are often epiphytized by species of *Melobesia*, *Fos-*

liella, Heteroderma, and *Dermatolithon* that do not appear modified for their habitat. Some species in *Clathromorphum, Mesophyllum,* and *Synarthrophyton* are button-like crusts apparently growing obligately on corallinaceous and noncorallinaceous hosts.

Some extensively modified coralline algae are parasitic on other coralline algae. Additionally, *Schmitziella* grows between cell wall layers of the green alga *Cladophora*. *Kvaleya* and *Ezo* are unpigmented examples of adelphoparasites, the former on *Leptophytum* and the latter on *Lithophyllum*. The vegetative filaments of *Choreonema* grow among intergenicular filaments of articulated coralline algae and only the conceptacles are produced externally.

Chapter 4

STRUCTURE OF ARTICULATED CORALLINE ALGAE

INTRODUCTION

Over the years, three types of branching coralline algae evolved soft tissues interspersed among the hard parts. Although the old grouping of the Corallinaceae into two formal groups, the nonarticulated and the articulated, is not now advocated, it is convenient to treat them in separate chapters. The three lines of evolution leading to articulated entities resulted in the development of three different types of genicula. The three recognized subfamilies will be treated in turn.

CORALLINOIDEAE

This subfamily contains species having intergenicula with several tiers of medullary cells, all of about the same height, and single-tiered genicula. These intergenicula and genicula are organized into branching fronds which arise from crustose or rhizomatous holdfasts.

A branch consists of a system of one or more intergenicula arising from another intergeniculum. A frond consists of a system of branches arising at one point from a holdfast. In most Corallinoideae many branches make up a frond; exceptions are *Yamadaea* where the fronds consist of only one or two intergenicula and *Chiharaea* with fewer than 15 intergenicula per frond (Figure 1). Frequently, several fronds grow together in a clump from a seemingly single base. Sometimes, in genera such as *Jania*, frond axes cannot be traced upward very far because they bifurcate into other axes equal in extent and bifurcating again (Figure 2).

INTERGENICULA

Intergenicular size and shape are often characteristic of certain genera and species, although recent evidence shows that environment influences intergenicular morphology.[58,235] The lowermost intergenicula in most fronds — as well as those in the upper parts in most species of *Jania, Bossiella cretacea*, and in plants growing in subdued light and little surge — are round in transection.[235] In most species however, the intergenicula in the upper parts of the main axes are mostly flat. In some taxa, such as some species of *Bossiella*, a midrib protrudes slightly, separating a pair of flat extensions called *wings*. If lateral extensions extend above the upper edges of intergenicular axes, they are called *lobes*, characteristic the genus (Figure 2). If development is not interfered with, the wings or lobes characteristic of certain taxa are repeatedly expressed in successive intergenicula. Most intergenicula are broader distally than proximally; note especially the cuneate axial intergenicula of *Corallina* (Figure 2).

In the upper parts of fronds in some species of *Corallina*, intergenicula may become excessively branched and irregular in shape without the development of subtending genicula (Figure 2). It appears that the usual segmentation and genicula production becomes disrupted so that these intergenicula possibly represent several intergenicula that have become congenitally fused (see plate 9, Figures 7-20, of *Corallina pinnatifolia* var. *digitata* in Dawson.[113])

The smallest fronds are those of *Yamadaea* which consist of one to two erect intergenicula not more than 3 mm long.[389] Extreme delicacy occurs in *Jania capillacea* where the intergenicula are less than 100 μm in diameter.[432] Intergenicula are smaller

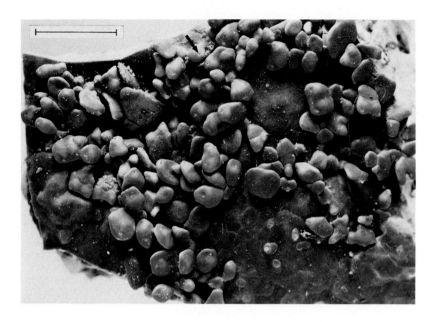

FIGURE 1. *Chiharaea bodegensis*. The basal crust is extensive and some of the intergenicula contain conceptacles in which the pores are visible (arrow). Scale = 5 mm. (Johansen, H. W., *Phycologia*, 6(1), 51, 1966. With permission.)

than usual and tend to be barrel-shaped when species (e.g., *Corallina officinalis, Cheilosporum sagittatum*) are grown in the artificial environment of a laboratory,[58,235] and in deep water.[235] On the other hand, the fronds of *Calliarthron cheilosporioides* sometimes grow to a height of 30 cm off the coast of California, and consist of flat intergenicula up to 3 mm long and 6 mm wide.[223]

The intergenicula in some species are dorsi-ventral. In *Chiharaea bodegensis* fronds consist of no more than 14 intergenicula in branching sprays that are appressed to a crustose holdfast.[221] In this species the tetrasporangial conceptacle pores are on the upper intergenicular surfaces (Figure 1). A similar situation occurs in *Arthrocardia duthiae* and *Arthrocardia silvae* in which fronds do not tend much toward proneness, but nevertheless have excentric pores even though the conceptacles are located deep within the intergenicula.[226] In *Bossiella californica* ssp. *schmittii*, an entity occurring subtidally off the coast of western North America, the fronds often grow more or less horizontally from the sides of rocks. The large, flat intergenicula bear most or all conceptacles on the dorsal surface, some specimens having as many as 50 conceptacles per intergeniculum.[230] The ventral intergenicular surfaces tend to be slightly concave and pale in color.

Intergenicular shape is often influenced by the presence of conceptacles and, especially in sexual plants, the large number of conceptacles present on each intergeniculum may cause any characteristic shape to be lost, (e.g. *Calliarthron tuberculosum*).[223]

With precise synchrony apical meristems produce tiers of intergenicular cells in which the primary pit-connections form arching lines (Figure 3). At intervals of 5 to 20 tiers of intergenicular cells (or even as many as 70 in *Calliarthron tuberculosum*), a tier of genicular cells is produced. The number of tiers per intergeniculum and the height of the tiers is relatively constant within a species, but varies among the taxa. In *Corallina officinalis* about 14 tiers are produced between genicula, each tier about 80 μm long.[235] In *Calliarthron tuberculosum* as many as 70 tiers, each 50 to 75 μm tall, may form between genicula.[223,234] The other extreme occurs in some species of *Jania*

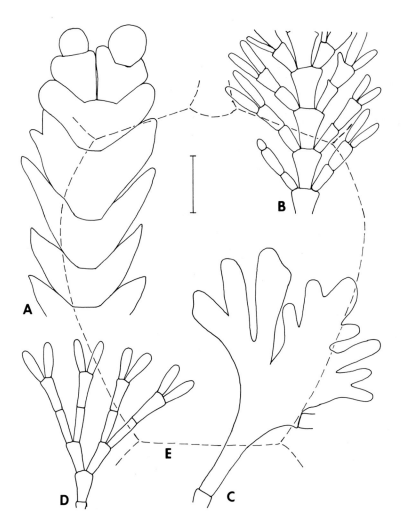

FIGURE 2. Magnified tracings of an assortment of intergenicula in the Corallinoideae. A. *Cheilosporum cultratum*. Note the dichotomous branching near apex. B. *Corallina officinalis*. Every intergeniculum in the main axis bears two lateral branches. C. *Corallina officinalis*. A single intergeniculum that is apparently a fusion product of several intergenicula. D. *Jania rubens*. The branching is dichotomous, but not all intergenicula branch. E. An intergeniculum of *Bossiella californica* ssp. *schmittii* is shown by the dashed outline. Scale = 1 mm.

and *Haliptilon*. In *J. rubens* there are 4 to 6 tiers per intergeniculum, each 100 to 170 μm tall, and in *H. squamatum* 5 to 9 tiers, each 90 to 120 μm tall.

In most species medullary cells are fairly straight, but Manza[294] noted that some plants from Central California had "filamentis medullaribus intergeniculorum flexuosis intertextisque" (Figure 3). This was his primary characteristic for establishing the genus *Calliarthron* in which he later[301] placed five species from California and three from Japan. Segawa verified this for *Cheilosporum yessoense*[383] and *Alatocladia modesta* (as *Cheilosporum anceps* var. *modesta*),[388] both taxa which Manza put into *Calliarthron*.

Cortical filaments are initiated when meristematic cells at the periphery of an apex cut off epithallial cells. The cells below the epithallial cells constitute an intercalary meristem, dividing to produce cortical cells below the apex while the primary meristem continues to be responsible for branch elongation. As the cortical filaments are grow-

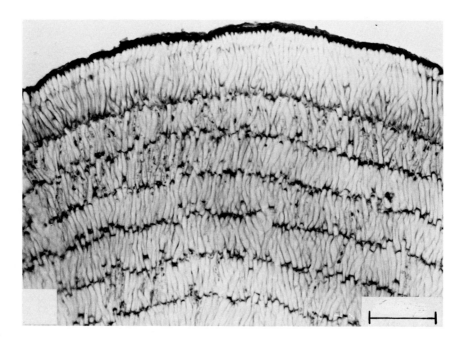

FIGURE 3. Longisection through apex of a branch of *Calliarthron cheilosporioides* in which the meristematic cells have recently divided. The medullary cells are interlacing, as is characteristic of *Calliarthron*. Scale = 100 μm. (Johansen, H. W., *Univ. Calif., Berkeley, Publ. Bot.*, 49, 1, 1969. With permission.)

ing, plastids form within the cells. The tips of coralline branches are white because they are uncoated by pigmented cortical tissue. Cortical thickness varies, being thin in some species of *Jania* and *Haliptilon*, and thicker in more robust species, such as *Calliarthron tuberculosum*.

GENICULA

Of prime importance in delimiting the articulated coralline algae from other entities are genicula, the groups of uncalcified cells separating intergenicula, rendering the branches flexible. Genicula have been studied by numerous workers and several facts that characterize corallinoidean genicula have been discovered:

1. Genicula consist of single tiers of cells (Figure 4).
2. For the most part genicular cells are uncalcified; however, the ends are calcified, projecting into adjoining intergenicula and attached to intergenicular medullary cells by primary pit-connections.
3. No cortical tissue is produced from genicula, that is, the genicular cells are unbranched.
4. The uncalcified walls are thick and stain intensely with numerous stains; their composition is unknown.
5. The cells are long (up to 0.5 mm) and thin (6-10 μm).
6. They are initiated at regular intervals by the medullary meristem.
7. Unlike intergenicular medullary cells, they elongate as they age.
8. In some taxa, e.g. *Calliarthron tuberculosum*, a developing geniculum is endogenous from the time a new intergenicular tier of cells is produced above it until the cortex ruptures so as to expose it (Figures 5,6).[223]

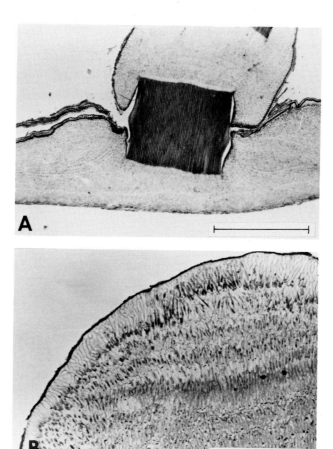

FIGURE 4. Geniula in *Calliarthron*. A. Section through a basal crust and the basal geniculum of a frond of *Calliarthron tuberculosum*. The lower surface is smooth because the plant grew on a glass slide in a *Macrocystis* forest at Point Lobos, Calif. Scale = 500 μm. B. A young geniculum of *Calliarthron* sp. showing that it is endogenous and that the cells may be distinguished from those of the intergenicular medulla. The genicular cells are elongating and the uncalcified parts of the walls are more stainable. Scale = 200 μm. (Johansen, H. W., *Univ. Calif., Berkeley, Publ. Bot.*, 49, 1, 1969. With permission.)

9. Decalcification apparently occurs when the cortex around an elongating geniculum ruptures (Figure 6).[223]
10. The flexibility of genicula is limited by calcified phlanges that grow up and down from adjoining intergenicula.

The walls of genicular cells are dense and stain readily with numerous reagents. As they elongate and locate farther below branch apices the staining characteristics change. Borowitzka and Vesk,[50] in an ultrastructural study of *Corallina officinalis* and *Corallina cuvieri* (= *Haliptilon cuvieri*), found that the amount of fibrillar material in the walls increases with age. Relatively mature genicular cells of *C. officinalis* have walls 0.7 to 1.0 μm thick made up of densely packed fibrillar material. The walls are bordered by a thin, dense, outer cuticle-like layer. The middle lamella lies between the

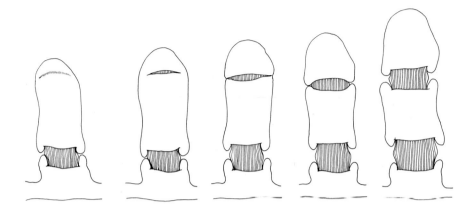

FIGURE 5. Diagrams of stages in the growth of genicula in *Calliarthron tuberculosum*. The young geniculum is endogenous until cell elongation coupled with autolysis result in cortical rupture. (Johansen, H. W., *Univ. Calif., Berkeley, Publ. Bot.*, 49, 1, 1969. With permission.)

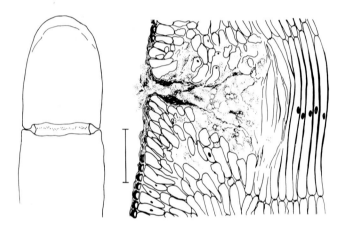

FIGURE 6. Cortex rupturing during growth of a geniculum in *Calliarthron tuberculosum*. These genicular cells are about 150 μm long; at maturity they would be up to 500 μm long. Scale = 50 μm. (Johansen, H. W., *Univ. Calif., Berkeley, Publ. Bot.*, 49, 1, 1969. With permission.)

walls of adjacent cells, is about 0.3 μm thick, and is filled with fibrils and amorphous material that stain differently from the walls proper.[50]

The relative lengths of intergenicular and genicular cells are such that in almost all taxa the latter are much longer. An exception is *Jania rubens* where both cell types are similar in length, being 100 to 170 μm and 90 to 160 μm long, respectively.

Genicula are remarkably uniform in structure throughout the Corallinoideae. At the same time, they differ markedly from genicula in the Amphiroideae and Metagoniolithoideae. They also serve as important features delimiting these three subfamilies. Apparently genicula evolved in corallinoidean ancestors as a method of raising and making flexible erect portions of plants. This would make for wide spore dispersal and would be a way of raising plants into the light above neighboring competitors. The genicula are an integral part of the fronds, as may be visualized when following development in Figure 5. Many questions may be posed about these unique structures, especially regarding developmental cytology. A modern study along the ideas of Yendo's[457] treatment of 75 years ago would be welcome.

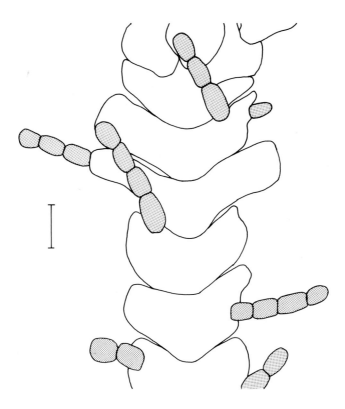

FIGURE 7. Magnified tracing of *Bossiella orbigniana* ssp. *dichotoma* from which grew secondary branches (shaded) when the plant was in culture for six weeks. Scale = 1 mm.

BRANCHING

For many years branching in articulated coralline algae has been described as dichotomous, pinnate, and in some taxa, palmate. Before analyzing branching in these algae, primary and secondary branches must be distinguished. Primary branches originate near the apices of axes, while the basic intergenicular form is still developing. Secondary branches appear later from intergenicula that may even be in lower parts of the fronds. These kinds of branches often appear to be the result of injury, changed orientation of the frond, or changed environmental conditions, such as when plants are placed in laboratory culture (Figure 7). Some taxa, such as *Jania prolifera* and *Cheilosporum proliferum*, have a propensity for producing secondary branches.

Whether an intergeniculum is producing 1, 2, or 3 surmounting intergenicula, the first visible indication of branching is when the apical meristem cuts off cells destined to become genicular. In delicate species, such as *Corallina officinalis*, genicular cells appear to elongate rapidly, hence the geniculum soon becomes visible between the new intergeniculum and the older one below. In coarse plants, such as *Calliarthron tuberculosum*, the young genicula remain concealed within branch apices (Figure 5), even as new tiers of intergenicular cells are produced above.[223] The calcified intergenicular tissue surrounding an elongating geniculum in *C. tuberculosum* breaks down only when it is several tiers below the apex (Figure 6). Branch buds become visible because, coupled with rapid elongation, the genicular cells do not produce cortical cells. Hence, there is a space devoid of cortex between a bud and its underlying intergeniculum (Figure 8).

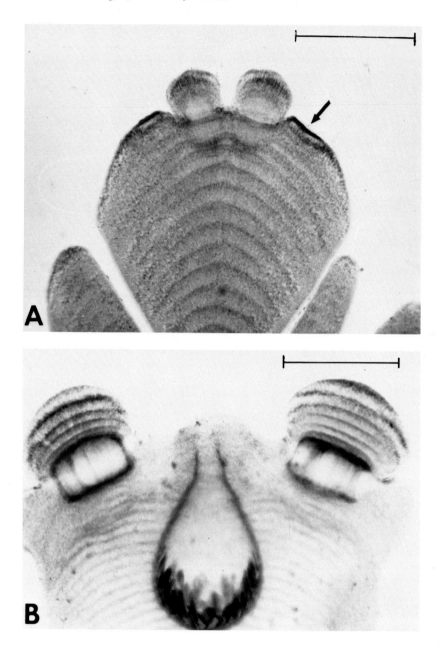

FIGURE 8. Dichotomous branching. A. *Cheilosporum*. Conceptacle primordia (arrow) are present on either side of the two branch buds. Scale = 500 μm. B. A fertile intergeniculum of *Arthrocardia*. In this genus branching is pinnate in the vegetative parts and dichotomous where conceptacles are present. The conceptacle is axial and located between the branch buds. Scale = 500 μm. (Johansen, H. W., *Univ. Calif., Berkeley, Publ. Bot.*, 49, 1, 1969. With permission.)

The first several intergenicula of a frond growing from a holdfast are often terete and unbranched, a single geniculum arising from each and, in turn, giving rise to other intergenicula and genicula in a linear series. Soon, the upper part of an intergeniculum loses its terete form, becoming slightly flat and broad. Then, instead of one, two branch buds, for example in *Jania* and *Cheilosporum* (Figure 8), or three, as in *Corallina* and some species of *Bossiella* (Figure 2), form from the primary meristem at the top of the intergeniculum. Development of these buds continues until each becomes an intergeniculum separated from the intergeniculum below by a geniculum. In *Jania* intergenicula in upper parts of the fronds may be more or less terete until branching occurs. Then the upper parts of a branching intergeniculum flatten and broaden, two branch buds arise, and the branches develop.

When three new intergenicula are produced in *Corallina* and *Arthrocardia* the central one grows more rapidly than the laterals and the main axis is perpetuated. Here branching usually occurs in successive axial intergenicula. In most corallinoidean plants the flattening and broadening of intergenicula about to branch occurs in the same plane and consequently the branches lie in the same plane. An exception is *Jania adherens*, which is often more or less decussate.[113] A form of this species from Japan was described as *J. decussato-dichotoma*.[455] The type of branching in *Jania* has been called *dichotomous* and that in *Corallina, pinnate*. Palmate branching occurs irregularly in some species of *Corallina* and *Arthrocardia*, especially in upper parts of the fronds. As many as seven or eight branches may arise from a single intergeniculum, but often the branches consist of only single intergenicula. Branching may also be influenced by the presence of conceptacles, for example in *Arthrocardia* (Figure 8), *Jania*, and *Haliptilon* where two branches instead of the usual three arise from fertile intergenicula. The branching patterns occurring in the Corallinoideae are given in Table 1.

The variation in form imposed by environmental factors has been hinted at by von Stosch,[418] Colthart and Johansen,[102] Johansen and Colthart,[235] and Mendoza,[317] but further research is needed here. In one study,[235] laboratory-grown fronds of *Corallina officinalis* from the western Atlantic differed from those collected from the shore: (1) intergenicula were shorter and narrower, (2) medullary cells were shorter, (3) genicula were shorter, and (4) cuticles were thicker. It remains to be seen what specific factors result in such modification.

The effect on morphology of altering the seawater medium could provide data on basic processes in coralline growth, but little work of this nature has been done. Von Stosch[420] showed that the successful growth of *Jania rubens* required media with unusually low phosphate concentrations, but this sensitivity was not evident in *Corallina officinalis*. Recently, Brown et al.[58] found a similar phosphate intolerance in *Cheilosporum sagittatum, Jania rubens,* and *Corallina officinalis* from southern Australia.

FROND INITIATION

The initiation of fronds from holdfasts in the Corallinoideae has been observed in species of *Corallina, Jania, Haliptilon,* and *Calliarthron* by Cabioch,[63] Johansen,[223] and Johansen and Austin.[234] In this subfamily a frond is attached to its base by a basal geniculum (Figure 4). As shown by Cabioch,[63] there are two ways in which fronds are produced. In *Corallina officinalis* and *C. elongata* (as *C. mediterranea*) spores germinate slowly into substantial crusts that may become as much as 5 mm in diameter before fronds appear several months later. These basal crusts resemble *Lithothamnium*, as do also the bases of *Calliarthron tuberculosum*.[223] In *C. officinalis* the hypothallia consist of two strata of filaments above which thick perithallia develop. The perithallial meristems give rise at irregular intervals to groups of cells which produce the basal genicula of incipient fronds.

In the second manner of frond initiation, as exhibited by *Jania rubens* and *Haliptilon squamatum*, four-celled sporelings give rise to tiny, weakly attached crusts or to no crusts at all. Whichever the case, fronds are initiated in a matter of weeks, certainly less than one month. As Cabioch[63] has pointed out, development in these plants may be an adaptation for being swept into and entangled with host plants where rapidly

Table 1
THE NUMBERS OF BRANCHES ARISING FROM VEGETATIVE AND REPRODUCTIVE INTERGENICULA IN THE CORALLINOIDEAE

Genera	Branching Vegetative	Branching Reproductive	Comments
Alatocladia	2,3	2	Variable, dense
Arthrocardia	3	2	Dense
Bossiella	2,3	2,3	Dense or sparse, depends on species
Calliarthron	2,3	2,3	Variable, dense
Cheilosporum	2	2	Sparse
Chiharaea	2,3	0	Variable
Corallina	3	0,(2)	Sexual conceptacles sometimes branched, dense
Haliptilon	2,3	2(0)	Conceptacles "antenniferous", no branches from male conceptacles
Jania	2	2(0)	Conceptacles "antenniferous", no branches from male conceptacles
Marginisporum	2,3	2,3	Variable, dense
Serraticardia	3	0,3	Dense
Yamadaea	0	0	Fronds unbranched

developing secondary crustose bases become established. These plants are usually epiphytic, whereas *Corallina officinalis* and *C. elongata* are usually epilithic. The differences in the mode of frond initiation is another feature which serves to taxonomically link *Jania* and *Haliptilon* and set them apart from *Corallina*. Fully grown fronds in epiphytic plants in these two species and in *Cheilosporum* are attached by a tangle of rhizomes made up of small barrel-shaped intergenicula (Figure 9).

Studies of the relatively robust *Calliarthron tuberculosum* growing on settling surfaces revealed that fronds were initiated within two months after germination.[223,234] The substantial nature of the holdfast indicate that development in these plants is similar to that in the slow-growing *Corallina*.

AMPHIROIDEAE

Although members of this subfamily superficially resemble those in the Corallino-

FIGURE 9. The lower part of *Cheilosporum* sp. showing the holdfast of tangled terete branches rather than a crust. Scale in millimeters. (Johansen, H. W., *Phycologia*, 15(2), 221, 1976. With permission.)

ideae by virtue of having fronds made up of intergenicula and genicula, the internal structure is so different that there is no question of the distinctiveness of the two groups. In retrospect, it is surprising that the Amphiroideae as well as the Metagoniolithoideae, were not segregated from the Corallinoideae until 1969.[223]

AMPHIROA

The fronds in *Amphiroa* are up to 30 cm long and comprised of many intergenicula (*A. ephedraea*). They can also be small and consist of extensive basal crusts and fronds consisting of only one or two intergenicula (*A. crustiformis*). The intergenicula are terete to flat, and in some species they are long relative to width (e.g., up to 20 times longer than broad in *A. fragilissima*).[432] In some species, such as *A. misakiensis*, the fronds are flat and form rosettes in which the intergenicula are more or less prone with upper surfaces convex (Figure 10).

Branching is usually dichotomous although sometimes it is irregular in aspect. In most species the fronds tend not to lie in one plane, but instead form small bushy clumps (Figure 10). The two genicula at a dichotomy are occasionally not located on the upper margin of the subtending intergeniculum, but rather the dichotomy is below them and is a feature of the intergeniculum itself (Figure 10). Secondary branching is prevalent in some species.

The anatomical features of *Amphiroa* are revealed in longitudinal sections through branch apices. The cortex and medulla are more clearly demarcated than in the Corallinoideae (Figure 10). There is also a greater tendency for cortices to thicken as intergenicula age, such as in *A. ephedraea* (Figure 11).[224] The most notable feature in longitudinal sections of branches are the varying heights of the precisely organized tiers of medullary cells (Figure 10). Although developmental studies of apical meristems in *Amphiroa* have not been made, it appears that tiers of short cells are produced following which cells in some of the tiers elongate, some to as much as 120 μm. This elongation takes place below a growing apex and occurs not at all in some tiers and to lesser extents in others. The possibility that sequential patterns of relative tier heights may

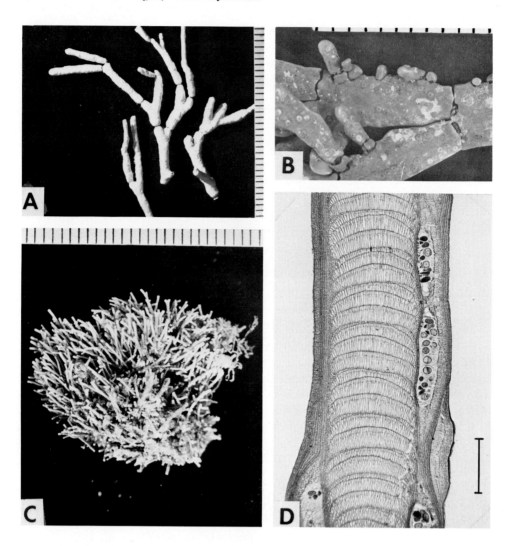

FIGURE 10. *Amphiroa*. A. *Amphiroa subcylindrica*. In some intergenicula the branching is below the point where the genicula arise. Scale in millimeters. B. *Amphiroa anceps*. Several secondary branches are present as are also numerous small conceptacles. Scale in millimeters. C. *Amphiroa valonioides*. In this delicate species each geniculum consists of only a single tier of cells. Scale in millimeters. D. Longisection of *Amphiroa ephedraea*. Note the sharp demarcation between medulla and cortex and the differing heights of the tiers of medullary cells. Scale = 300 μm. (Fig. 10D, Johansen, H. W., *J. Phycol.*, 4(4), 319, 1968. With permission.)

be diagnostic for some species seems not to hold, but this information is often given in descriptions.[175,197]

Other features that characterize *Amphiroa* medullae are secondary pit-connections and epithallial cells that are located above the apical meristems.[377,378] Hence, even at branch apices, the meristem is an intercalary one, unlike that in the Corallinoideae. The formation of this apical epithallus in *Amphiroa* should be investigated. Most reports reveal noncalcified wall layers *between* the epithallus and the underlying meristem.[377,378]

GENICULA

It appears that genicula in *Amphiroa* develop when groups of intergenicular cells

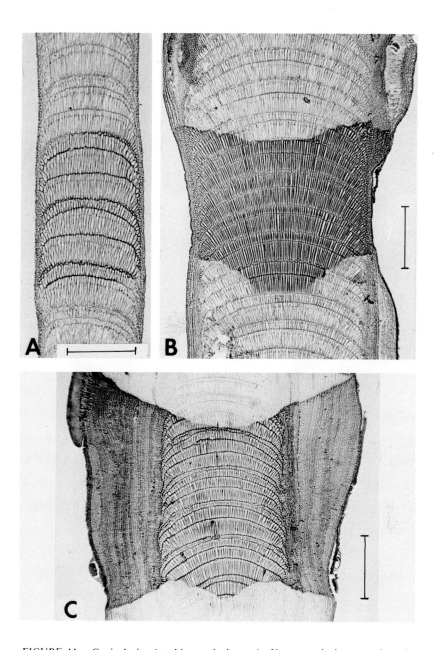

FIGURE 11. Genicula in *Amphiroa ephedraea*. A. Young geniculum near branch apex. B. Further from the apex. C. Near base of frond. Note progressive thickening of cortex. Scale 300 μm in the three figures. (Johansen, H. W., *J. Phycol.*, 5(2), 118, 1969. With permission.)

near branch tips progressively become decalcified. Thus, genicula can only be seen below the tip when the component cells have become decalcified enough to stain more intensely than the surrounding calcified cells (Figures 11, 12). In sectioned material, the first visible evidence of a geniculum appears near the center of a branch several hundred micrometers below its apex (Figure 11).[224] Unlike the Corallinoideae, there are no cell length differences between intergenicula and genicula. However, little is known about genicular initiation in *Amphiroa*. In species such as *A. rigida* initiation may be more than simply a decalcification of intergenicular cells.

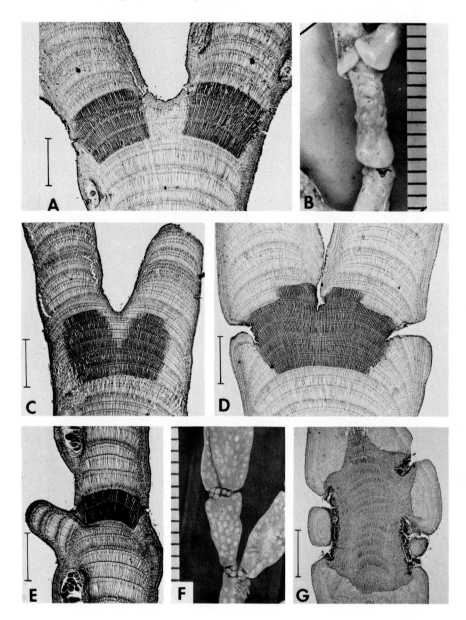

FIGURE 12. Type 1 genicula in *Amphiroa*. A. *Amphiroa beauvoisii*. The intergenicular cortex has cracked around the genicula. Scale = 300 μm. B. *Amphiroa capensis*. Intergenicular aprons cover the genicula. Scale in millimeters. C. *Amphiroa anceps*. Young genicula still endogenous. Scale = 400 μm. D. *Amphiroa capensis*. Genicula fully developed. Scale = 300 μm. E. *Amphiroa* sp. Genicula here consist of two cellular tiers. Note incipient secondary branch. Scale = 250 μm. F and G. *Amphiroa bowerbankii* showing minisegments. F. Scale in millimeters. G. Scale = 400 μm. (Johansen, H. W., *J. Phycol.*, 5(2), 118, 1969. With permission.)

According to Johansen,[224] decalcification in some species begins near the center of a branch and proceeds centrifugally until the surface is reached. In other species decalcification stops short of the branch surface (Figure 12). The first instance, where genicula become decalcified all the way through a branch, occurs in the robust *A. ephedraea* in southeastern Africa. This is the type II genicular development (Figure 13).[224] There is little to retard the bending of the relatively massive genicula, which may con-

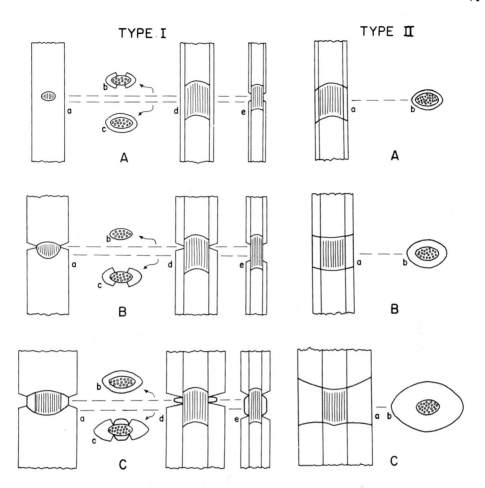

FIGURE 13. Diagrams of stages in genicular development in *Amphiroa*. Genicular medullary tissue is represented by longitudinal striations in lateral views and by small circles in transverse views. Genicular cortical tissue is shaded. Intergenicula are unmarked except that medullary tissue is separated from cortical tissue by lines. Type I diagrams are external views (a) transections (b and c) frontal sections (d) and sagittal sections (e) They show a stage after a cortical patch has sloughed (A) a stage after the calcified cortex has cracked all around geniculum (B) and an old stage after secondary cortical tissue has been produced (C). Type II diagrams are longisections (a) and transections (b). They show a young geniculum (A) a stage after some secondary growth (B) and a stage after extensive secondary growth (C). (Johansen, H. W., *J. Phycol.*, 5(2), 118, 1969. With permission.)

sist of as many as 25 medullary tiers and the overlying cortex (Figure 11). These genicula continue to grow slowly as more and more tissue becomes decalcified until a large amount of the tissue at the base of the frond is genicular. In living plants of *A. ephedraea* the genicula are conspicuous black bands 0.5 to 1 mm long separating the intergenicula which are 3 to 6 mm long (Figure 11).

In another type of genicular development (Type I),[224] characteristic of most species of *Amphiroa* with flat intergenicula, decalcification does not proceed to the branch surface. Instead, when a young endogenous geniculum is 2 to 4 mm below a branch apex, the calcified covering cracks and partly breaks off so as to expose the underlying geniculum (Figure 12). The dark genicular tissue then becomes visible as a spot on the flat branch surface. Further growth, and probably the bending pressures of wave action, result in more calcified tissue breaking away. This type of development is best seen in flat branches such as those of *A. anceps* (Figure 2, Chapter 1). Unlike Type II

Table 2
GENICULAR CHARACTERISTICS IN SOME COMMON SPECIES OF *AMPHIROA*

1 medullary tier	*A. valonioides*
2 medullary tiers, oblique transverse walls	*A. rigida*
3 — 5 medullary tiers	*A. beauvoisii*
5 or more medullary tiers; type II development	*A. ephedraea*
5 or more medullary tiers, type I development	*A. anceps*
5 or more medullary tiers, type I development; minisegments	*A. bowerbankii*
Genicula bracketed by swollen ends of intergenicula in older parts of fronds	*A. fragilissima*
Genicula covered by pronounced "apron" of calcified tissue produced by intergenicula above	*A. capensis*

(*A. ephedraea*), the branch surface is not continuous over the genicula, instead being interrupted by cracked calcified cortices.

In most species of *Amphiroa*, there are two or more tiers of medullary cells (plus some cortical tissue) in a geniculum, with at least one species, *A. valonioides*, having only a single tier. As Weber-van Bosse[444] pointed out, genicular anatomy is useful in distinguishing the species (Table 2). In some entities, the genicular tiers are wholly uncalcified (e.g., *A. rigida*). In others, decalcification proceeds from a central area without regard for tier, or even cell boundaries (Figure 11). For example, at the upper edge of a geniculum, a tier may be composed of both calcified and uncalcified cells. The demarcation between calcified and uncalcified tissue may even pass through cells. Such a cell would be partly intergenicular and partly genicular.

Unusual features associated with genicula occur in some species of *Amphiroa*. Small, irregularly-shaped pads of calcified tissue remain intact on genicular surfaces in *A. bowerbankii* (Figure 12). They have been termed "minisegments".[224] As genicula develop in *A. capensis*, overlying aprons grow down from the intergenicula above (Figure 12). *Amphiroa fragilissima* is characterized by havin intergenicula that are bulbous where they join the genicula.

LITHOTHRIX

Although belonging to the Amphiroideae, *Lithothrix* is so different from *Amphiroa* that it warrants separate treatment. The differences are not at the cytological level (e.g., both genera have secondary pit-connections and so on), but at the level of tissue development, such as the formation of genicula. Primary sources for information on the vegetative anatomy of *Lithothrix* are Ganesan and Desikachary,[180] Cabioch,[76] Gitting,[187] and Borowitzka and Vesk.[50]

Lithothrix, containing the single species *L. aspergillum*, is unique in that the intergenicula are made up of only single tiers of medullary cells and associated cortical and epithallial tissues. The branches are terete or slightly flattened and the intergenicula

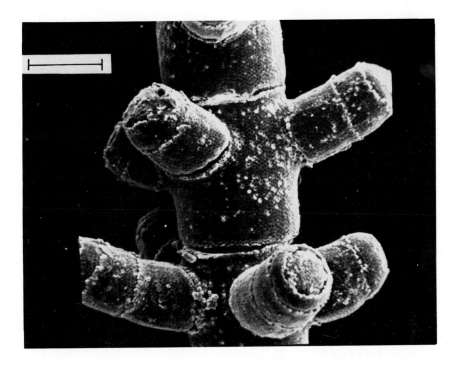

FIGURE 14. *Lithothrix aspergillum*. Part of branch showing axial intergenicula and several short lateral branches. Scale = 200 μm. Courtesy B. T. Gittins.

are less than 1 mm long (Figure 14). Branching is sparsely dichotomous, and irregularly distributed lateral branches are frequently abundant. The plants are easy to recognize in the field. It is restricted to the eastern Pacific from British Columbia to Baja California.[1]

The basal crust anatomically resembles *Dermatolithon*[76,188] and probably the articulated plants evolved from *Dermatolithon*-like ancestors. It is the fronds that are unusual.

The initiation of tiers of short medullary cells at branch apices is probably the same as in *Amphiroa*. Immediately after formation, the cells in alternating tiers commence elongating from about 10 μm up to 600 μm.[236] The result is that as apical growth continues, the bulk of the branch below consists of short-celled tiers alternating with tiers of cells which are progressively elongating (Figure 15). The short-celled tiers constitute intergenicular medullae whereas the long-celled tiers are genicular. Specifically, it is the lower or proximal parts of the long cells which are uncalcified and thus may be considered genicular (Figure 15).

In an ultrastructural study Borowitzka and Vesk[50a] described genicular cells as having large vacuoles and progressively more intensely staining cell walls. At full development the walls are about 2 μm thick and consist of concentric layers of fibrils 2 nm in diameter. Development involves the continued deposition of these fibrils within the cell walls. The cells are separated by intercellular spaces wherein occur fibrils (probably mucilage). In the upper one third of the genicular cells the intercellular spaces and the cell walls contain calcium carbonate.

During growth the intergenicular cells do not elongate, but produce new cells by a marginal meristem. These new cells are cortical, growing downward to form a flange of tissue covering the subtending elongating genicular tier. In surface view, it is these flanges that are interpreted as intergenicula. Note in Figure 15 how the intergenicula are progressively longer below the branch apex.

74 Coralline Algae, A First Synthesis

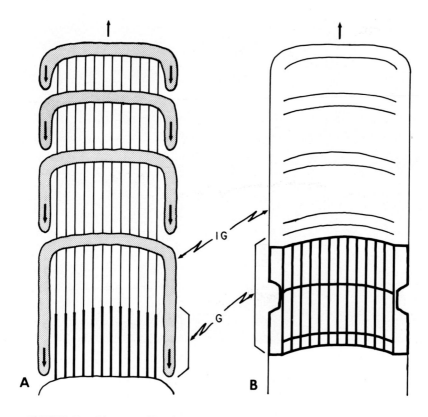

FIGURE 15. Diagrams of longisections of branch apices. The arrows indicate directions of growth. Uncalcified walls in genicular cells are thickened. A. *Lithothrix*. Intergenicular tissue is shaded. B. *Amphiroa*. IG = intergenicula; G = genicula.

METAGONIOLITHOIDEAE

Metagoniolithon, the only genus in the Metagoniolithoideae, contains three species restricted to western and southern Australia.[144] From descriptions in Segawa,[388] Johansen,[223] Ganesan,[178] Cabioch,[76] and, especially, Ducker,[144] it becomes clear that this group is distinctive, particularly as regards the vegetative anatomy of genicula. Most of the information on vegetative features given below is from Ducker.[144]

The fronds may be dense and bushy and as much as 30 cm long in *M. stelliferum*. The bushiness stems from the numerous branches arising from the genicula, with as many as 20 branches (but usually fewer) emanating from each geniculum (Figure 2, Chapter 1).

The meristematic abilities of genicula in *Metagoniolithon* are unique. Distinctive meristematic cells terminate medullary filaments at branch apices; epithallial cells are absent. The distinctiveness derives from the fact that the meristematic cells are free from one another, although the medullary cells derived from the meristem comprise the usual compact tissue (Figure 16). Most conspicuous is a cap of mucilaginous material overlying each apex; these caps may become as much as 250 μm thick in *M. radiatum*[144] and are reminiscent of angiosperm root caps, except that in *Metagoniolithon* they are acellular (Figure 16). Below this cap the meristematic cells divide synchronously during growth of the intergenicula. Shortly after the transformation of an apex to the production of genicular cells, it divides into two apices of approximately equal proportions (Figure 16). According to Ducker[144] this is universal in *Metagoniol-*

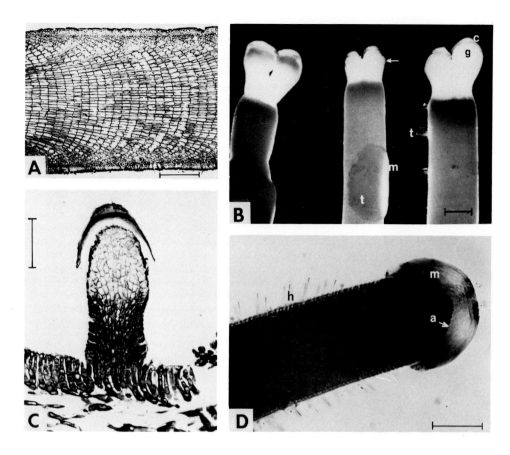

FIGURE 16. *Metagoniolithon*. A. Longisection through intergeniculum showing regular arrangement of medullary tiers of cells. Scale = 250 μm. B. Dichotomous branching of genicula in *M. radiatum*. Scale = 400 μm. C. Longisection of branch arising from geniculum in which intergeniculum (pale cells) and apical cap have been produced. Scale = 100 μm. D. Apex of branch *M. stelliferum* showing hairs (h), mucilaginous cap (m), and apical cells (a). Scale = 200 μm. (Fig. 16B, Ducker, S. C., *Aust. J. Bot.*, 27(x), 67, 1979; Fig. 16C, Johansen, H. W., *Univ. Calif., Berkeley, Publ. Bot.*, 49, 1, 1969; Fig. D, Ducker, S. C., *Aust. J. Bot.*, 27(x), 67, 1979. With permission.)

ithon. Growth continues and eventually the apices revert to the production of intergenicular tissues again.

The early development of genicula is not closely intergrated with intergenicular development, such as it is in the other two articulated subfamilies. Instead, in *Metagoniolithon* the apices clearly alternate between producing intergenicular and genicular tissues. Moreover, the meristematic activities of genicula continue while branch growth continues above them. This is manifested in their continued elongation until they become even longer than the intergenicula in *M. stelliferum*. It is also manifested in the production of lateral branches (Figure 16) that arise from the genicula rather than from intergenicula as in other articulated coralline algae. These lateral branches are the "false whorls" described by Ducker[144] and tend to obscure the primary dichotomous branching in all taxa except *M. chara* var. *dichotomum* in which lateral branches are not formed. Single "false whorls" develop from genicula in *M. chara* var. *chara*, and *M. radiatum* whereas several occur per geniculum in *M. stelliferum*.

Thus the genicula in *Metagoniolithon* are more prominent than in other articulated coralline algae. They are made up of medullary, cortical, and epithallial tissues. The cells are not organized into tiers, as they are in the integenicula. Unlike genicula in other coralline algae, cell fusions are present in genicular tissues.[144]

Anatomically, intergenicular medullary, cortical, and epithallial tissues are similar to corallinoidean entities, such as *Corallina*. The medullary cells are in tiers because of the synchronous divisions of apical meristematic cells (Figure 16). The cells are up to 60 μm long, but usually they are shorter and contrast with the longer medullary cells in the Corallinoideae. Trichocytes are usually present, occurring in bands near intergenicular apices in *M. chara,* whereas in *M. stellifera* they are scattered.[144]

Sheaths (or flanges or collars) grow from the intergenicula and partly cover adjoining genicula. In the variable species *M. stellifera,* these sheaths extend for considerable distances over genicula that may be so long that much of a frond axis is genicular rather than intergenicular.

Ducker[144] noted that the fronds arise from crustose holdfasts that have a cellular arrangement very much like that in the nonarticulated *Neooniolithon*. The interesting manner in which the holdfasts of epiphytic species of *Metagoniolithon* grow on plants of the nonarticulated *Heteroderma,* which in turn grow on benthic noncalcareous plants, such as the seagrass *Amphibolus,* is treated later.

SUMMARY

Three subfamilies of articulated coralline algae contain plants made up of holdfasts from which arise branching fronds of calcified intergenicula separated from one another by uncalcified genicula.

The Corallinoideae contains a genus (*Yamadaea*) with extensive holdfasts and fronds of only one or two intergenicula, whereas most genera contain plants made up of hundreds of intergenicula.

Intergenicula may be terete or they may be flat and endowed with wings, such as in *Bossiella,* or lobes, such as in *Cheilosporum.* To some extent intergenicular shape varies depending on environment, but not much is known about this. Medullary filaments are flexuous rather than straight in some genera.

Corallinoidean genicula consist of single tiers of long medullary cells. For most of their length these cells are thick walled, but uncalcified. The ends of the cells are calcified and attached to intergenicular medullary cells.

Primary branches arise at frond apices and secondary branches anywhere on intergenicular surfaces. Primary branching usually results in either two (dichotomous), or three (pinnate), branches from a branching intergeniculum. In some taxa branching may be pinnate in vegetative parts and dichotomous in fertile parts (e.g., *Arthrocardia*).

Holdfasts are crustose and long-lasting in most genera, such as *Corallina* and *Calliarthron.* However, in some epiphytic species they are only briefly crustose, and on contact with a host, rhizomatous branches grow from the lower parts of the fronds and anchor them.

The Amphiroideae contains two genera, *Amphiroa* and *Lithothrix*. These plants have secondary pit-connections and other Lithophylloideae-like characteristics.

In *Amphiroa* the medullary and cortical tissue is sharply demarcated. The most important feature of this genus is the medullary tiers of cells that are of differing heights. Epithallial cells cover branch apices as well as intergenicular surfaces. Some species have large fronds 30 cm tall and others have reduced fronds consisting of only a few intergenicula.

Genicula in most species of *Amphiroa* consist of several tiers of uncalcified medullary cells and usually, disrupted cortical tissue. Genicula are often similar to intergenicula except for the lack of calcium carbonate.

The monotypic genus *Lithothrix* has one-tiered intergenicula and one-tiered genicula, the latter consisting of very long cells. The short-celled intergenicula produce extensive cortices that grow downward and cover the genicula.

Three species in one genus are placed in the subfamily Metagoniolithoideae. Several features characterize these plants: (1) many-celled genicula from which several branches arise, (2) mucilaginous caps surmounting the branch apices, (3) noncontiguous meristematic cells at the branch apices, and (4) the initially dichotomous branching of apices producing genicular tissue.

Chapter 5

REPRODUCTION

INTRODUCTION

The tremendous amounts of coralline algae, and their presence for millions of years, reflects their remarkable ability to reproduce, both vegetatively and by means of specialized reproductive structures characteristic of the higher red algae. Plants usually bear conceptacles containing tetraspores (or bispores), carpospores or spermatia. Other plants growing in certain habitats only rarely form conceptacles, instead relying on a slow but marked ability to grow vegetatively (e.g., rhodoliths).

Except for the first part, this chapter will be devoted to the formation of conceptacles and the development of their contents. Numerous studies have been made of conceptacles in coralline algae, and they have provided a great deal of information that is useful in understanding phylogeny and classification as well as developmental processes.

VEGETATIVE REPRODUCTION

Although little data exist, coralline algae are obviously readily capable of vegetative reproduction. When articulated fronds are broken into pieces and cultured, they produce new fronds prolifically.[223] Fronds that remain in contact with hard surfaces produce crustose bases (attachment discs) from which new fronds arise. This phenomenon has been recorded for *Jania rubens*,[365] *Calliarthron tuberculosum*,[223] and several other species. Gittins[185] and Gittins and Dixon[188] found that intertidal turfs of *Lithothrix aspergillum* were maintained by extensive branching and the secondary attachment of branches with other branches and other algae. Deciduous propagules (Figure 1), probably functioning as dispersal units, may be produced in the upper parts of fronds of *Jania capillacea* from Pacific Mexico.[113] These three-pronged fragments form attachent discs when in contact with hard surfaces. They resemble the propagules in *Sphacelaria* (Phaeophyta).

Pieces broken from nonarticulated coralline algae continue to grow, if conditions are appropriate. Probably the great banks of marl (see Chapter 9) result when unattached thallia grow and fragment.

Suneson[423] suggested that germination experiments would be required to ascertain whether or not contracted protoplasts of enlarged marginal cells in *Fosliella limitata* (as *Melobesia limitata*) were, in fact, vegetative propagation organs. He noted that one of the two protoplasts per cell seemed to escape through slits in the cell walls. Hollenberg[208] observed multicellular "discoid bodies" in various stages of development growing from megacells of *Fosliella paschalis* from Southern California (Figure 1). Similar structures have been described in other species of *Fosliella* (as *Melobesia callithamnioides*) from the Gulf of Naples.[412] These bodies, which are stipitate or fan-shaped and, for the Southern California forms, up to 65 μm broad, may be vegetative dispersal units.

CONCEPTACLES

Reproductive cells reach maturity within conceptacles in all extant coralline algae except the specialized endophyte *Schmitziella*. Although conceptacles originate in various ways, when they are fully developed each consists of a fairly well-defined roofed

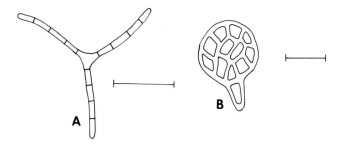

FIGURE 1. Vegetative reproduction. A. A deciduous propagule of *Jania capillacea* as observed by Dawson.[124] Scale = 1 mm. B. A possible dispersal unit produced by *Fosliella paschalis* as observed by Hollenberg.[208] Scale = 30 μm.

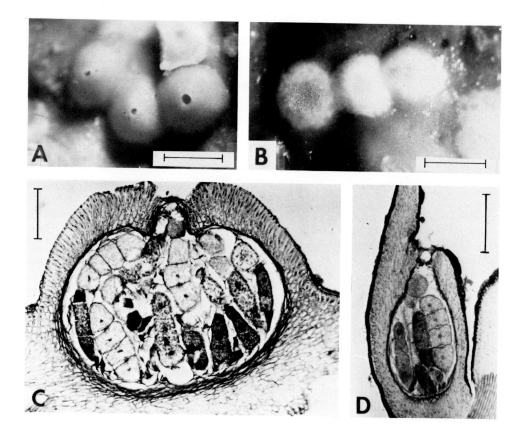

FIGURE 2. Tetrasporangial conceptacles. A. Single-pored conceptacles in *Lithoporella melobesioides*. Scale = 1 mm. Courtesy T. Masaki. B. Multi-pored conceptacles in *Lithothamnium pacificum*. Scale = 0.5 mm. Courtesy T. Masaki. C. Sectioned conceptacle of *Bossiella* sp. Scale = 100 μm. D. Sectioned conceptacle of *Cheilosporum cultratum*. Scale = 300 μm. Fig. 2C, Johansen, H. W., *Phycologia*, 15(2), 221, 1976; Fig. 2D, Johansen, H. W., *J. S. Aft. Bot.*, 43(3), 163, 1977. With permission.)

chamber opening to the outside by one or several pores (Figure 2). The open space between the pore and chamber is the *canal*. Uniporate conceptacles occur in all taxa except asexual plants of the Melobesioideae, where there may be as many as 60 or more pores in a conceptacle roof (e.g., *Lithothamnium*)[309] Asexual conceptacles in *Sporolithon* contain single tetrasporangia (Figure 5, Chapter 1), and groups of them have sometimes been called sori or nemathecia.[24]

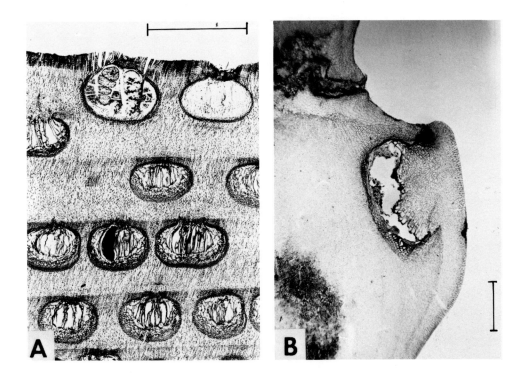

FIGURE 3. Conceptacle burial. Tetrasporangial conceptacles in *Clathromorphum nereostratum* that have been buried by vertical growth. Scale = 0.5 mm. Old conceptacle in *Calliarthron tuberulosum* invaded by secondary tissues. Scale = 250 μm. (Fig. 3A, Lebednik, P. A., *Syesis*, 9(x), 59, 1977; Fig. 3B, Johansen, H. W., *Univ. Calif., Berkeley, Publ. Bot.*, 49, 1, 1969. With permission.)

Probably conceptacles originated long ago as adaptations stemming from advantages in having protection for reproductive cells. The absence of fossil conceptacles in specimens from the Paleozoic[451] suggests that reproductive structures grew in superficial sori or nemathecia, perhaps similar to those in extant Peyssonelliaceae.[168] The earliest fossil coralline algae containing conceptacles are from the Jurassic Period.[450]

Conceptacles have also evolved in the noncorallinaceous but closely related Hildenbrandiaceae where they may be recognized more easily as sunken (depressed) sori.[1,253] In *Schmitziella,* the only coralline genus lacking conceptacles, bisporangia develop in groups between the cell wall layers of the green alga *Cladophora*.[424] The outer wall layer of the host serves as the roof of a chamber containing the sporangia. Spore release is through a pore forming in the cell wall layers above the sporangia.

Conceptacle development gives good clues to the evolutionary routes followed by various Corallinaceae. Furthermore, there are conspicuous differences between conceptacles in sexual and asexual plants, sometimes even in the same species (e.g., in the Melobesioideae). In *Clathromorphum* even the male and female conceptacles differ markedly in development.[262]

In nonarticulated coralline algae, conceptacles are located on the surface of crusts or on protuberances. Depending on the species, reproductive type, and the relative amount of surrounding vegetative growth, they may bulge conspicuously, or be flush with, or even sunken below the thallus surface. In thick plants, conceptacles may become completely buried by subsequent vegetative growth (Figure 3). Conceptacles originate behind growing margins in crustose species, apparently after a certain state of development or an environmental condition has been attained. In crusts of limited

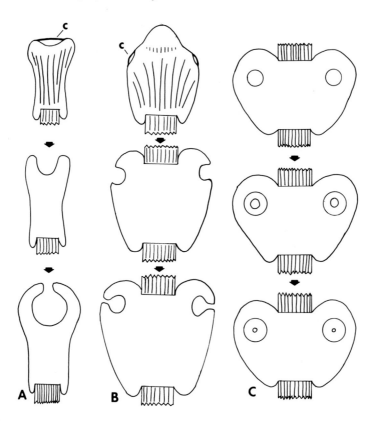

FIGURE 4. Diagrams of conceptacle types in articulated coralline algae. In the earliest stages (uppermost) the conceptacle primordia are overlain by caps (c). A. Axial conceptacles, where the primordia are in direct line with the longitudinal axis of the branch. B. Marginal conceptacles, where the primordia are on the shoulders or margins of intergenicula near branch apices. C. Lateral conceptacles, where the primordia are in cortical tissue on intergenicular surfaces. (Johansen, H. W., *Univ. Calif., Berkeley, Publ. Bot.,* 49, 1, 1969. With permission.)

expansion, such as in *Clathromorphum reclinatum*, the conceptacles may be in concentric rings as waves of them are periodically produced.[262] They originate when groups of perithallial meristematic cells differentiate into conceptacle primordia. In *Phymatolithon* and *Leptophytum*, conceptacles originate below the perithallial meristem.[13]

In articulated coralline algae conceptacles originate in medullary tissue at or near branch apices or in cortical tissue on intergenicular surfaces.[223] An unusual situation exists in two species of *Amphiroa, A. crustiformis*[119] and *A. currae*,[179] where conceptacles occur in the basal crusts as well as in intergenicula. There are two types of conceptacles originating in medullary tissue. They are both intimately associated with primary growth. *Axial conceptacles* originate in branch apices so as to usurp the straight line growth of the branch axis (Figure 4). In some taxa (e.g., *Jania*) having axial conceptacles, two branches may grow from a conceptacle roof, but neither one is in line with the subtending branch. In the second type of conceptacle originating in medullary tissue, the *marginal conceptacles*, origin is in the shoulders or lobes of intergenicula near branch apices. Here the branches continue to grow in their original direction (Figure 4). *Cortical conceptacles*, sometimes called lateral conceptacles, originate in cortices in intergenicula below the branch apices (Figure 4). These kinds of conceptacles may even form secondary crops in older parts of fronds. Conceptacle types in the corallinoidean genera are given in Table 1.

Table 1
GENERA OF THE CORALLINOIDEAE WITH INFORMATION ON CONCEPTACULAR CHARACTERISTICS

Genera	Conceptacle types[a]	Comments
Alatocladia	A, M	T[b] intergenicula sometimes branching. C and M unknown
Arthrocardia	A	Fertile intergenicula branching
Bossiella	C	1 — 50 or more conceptacles per intergeniculum
Calliarthron	M, C	
Cheilosporum	M	Fertile intergenicula not branching
Chiharaea	A (M)	Conceptacles are rarely clearly A or M
Corallina	A (PL)[c]	T and M intergenicula usually not branching; branches frequent in C intergenicula; PL especially in northern plants
Haliptilon	A	T and C intergenicula branching; M intergenicula not branching
Jania	A	T and C intergenicula branching; M intergenicula not branching
Marginosporum	M, C	C conceptacles sometimes appearing marginal after maturity
Serraticardia	A, C	Fertile intergenicula with A not branching
Yamadaea	A	Fertile intergenicula not branching

[a] Conceptacle types: A = axial; M = marginal; C = cortical.
[b] Fertile intergenicula: T = tetrasporangia; C = carposporangia; M = male.
[c] PL = pseudolateral conceptacles.

Modified axial conceptacles in which the branch axes and the subtending genicula abort are known as *pseudolateral conceptacles*.[223] They are homologous with secondary branches containing single axial conceptacles. Pseudolateral conceptacles resemble cortical conceptacles except that they protrude more. In *Corallina officinalis* they occur especially in northern specimens and in sexual plants. Their presence in *Corallina* has caused taxonomic confusion, and they are apparently not useful in classification. The genus *Joculator* was erected largely because of pseudolateral conceptacles in the type species *J. pinnatifolius*.[294] The genus is now merged with *Corallina*.

Mature conceptacles contain tetrasporangia (or bisporangia), spermatia, or carposporangia (the latter having developed following fertilization within female conceptacles). Typically these reproductive types occur on different plants. However, a few species are monoecious, (e.g., *Jania rubens* and *Serraticardia macmillanii*.)[226] In some species the distinction between dioeciousness and monoeciousness is not clear cut. For example, gametophytes of *Synarthrophyton patena* are either monoecious or dioecious, and occasional conceptacles are even hermaphroditic and contain spermatia as well as carpogonia,[435a] a phenomenon previously recorded for *Phymatolithon lenormandii*.[8]

The first visible evidence of conceptacle initiation is an increased secretion of cuticular material in the area where the conceptacle will form. This material is white when viewed under low magnification, but stains intensely with a variety of stains. Secretion is by the apices of meristematic cells and, if the meristem is intercalary, the epithallial cells are pushed upward by the deposits.[230] Hence, the cuticular material is not an addition to the cuticle that is already present over the epithallial cells, but a new secretion of the meristematic cells (Figure 5). In branch apices where epithallial cells are absent, the secretions increase the thickness of the preexisting cuticle. A thickened cuticle plus any associated cells overlying an incipient conceptacle is called a *conceptacular cap*. It probably serves to protect the young reproductive cells which will develop below. Conceptacular caps consisting only of secreted mucilage, may be as much as 100 μm thick, such as in the marginal conceptacles of *Calliarthron tuberculosum*.[223] In *Phymatolithon* and *Leptophytum* conceptacle origin is below the perithallial meristem in a newly formed endogenous meristem and the cap contains perithallial as well as epithallial cells. The occurrence of conceptacular caps in some genera, such as *Amphiroa* and *Metagoniolithon,* has not been clearly documented.

TETRASPORANGIAL CONCEPTACLES

There are four types of development of tetrasporangial (and bisporangial) conceptacles. These types are segregated by: the participation of filaments in forming the conceptacle roof, the way in which the chamber is formed, and the manner in which the exit pore(s) is formed.

1. Uniporate; *roof* formed of filaments overarching from around the fertile area; *chamber* underneath overarching roof; *canal and pore* is space surrounded by ends of roof filaments (Figures 6, 7).
2. Uniporate; *roof* formed of filaments which have developed from within the fertile area; *chamber* formed by breakdown of cavity cells among the reproductive cells; *canal and pore* formed (apparently) by local tissue breakdown (Figures 8, 9).
3. Multiporate; *roof* formed of filaments which develop from within the fertile area; *chamber* formed by breakdown of cavity cells among the reproductive cells; *canal and pore* formed in space occupied by a gelatinous plug, each plug forming above a tetrasporangium (Figures 10, 11).
4. Uniporate; *roof* formed of filaments which develop among crowded conceptacles; *chamber* is that space occupied by a single tetrasporangium; *canal and pore* formed as in number three, that is, in space occupied by a gelatinous plug (Figure 5, Chapter 1).

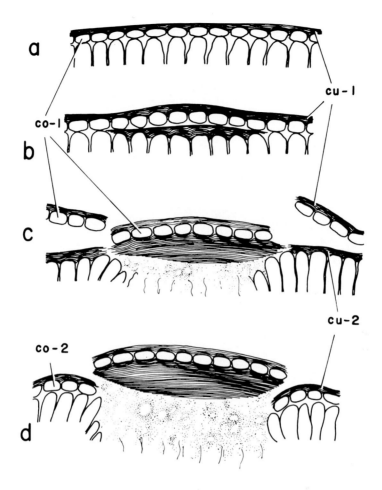

FIGURE 5. Diagram showing early stages in development in lateral conceptacles of *Bossiella californica* ssp. *schmittii*. A. Before initiation, with cuticle (cu-1) and epithallus (co-1). B. Secretion of conceptacle cap by meristematic cells below co-1. C. Dissolution of meristematic cells below cap and sloughing off of co-1 around cap and secretion of new cuticle (cu-2). D. Sloughing off of cap and the cutting off of a new layer of cover cells (co-2). (Johansen, H. W., *J. Phycol.*, 9(2), 141, 1973. With permission.)

These four types of conceptacles developed in different evolutionary lines and indicate taxonomic groupings as shown in Table 4, Chapter 1. Generally, most taxonomic usefulness at the subfamily level is gained by understanding development in tetrasporangial conceptacles rather than sexual conceptacles. Because of the distinctiveness, each type of conceptacle development will be treated in turn.

Type 1 (Figure 6) — Following secretion of a cap, the meristematic cells of the disc beneath apparently undergo partial autolysis and a space filled with cellular debris comes to separate the cap and these cells.[223,230] The separation and cellular destruction have been noted in certain species of *Bossiella*, *Calliarthron*, and *Corallina*. Many questions regarding this autolysis need to be answered. There is a transformation of all meristematic cells under the center of the cap into reproductive initials with the possible exception of some cells that grow into paraphyses.[389]

Subsequent development involves the origin and maturation of reproductive cells (treated later) and the growth of surrounding filaments to form the conceptacle. These filaments grow upward and then arch centripetally toward the center of the disc.

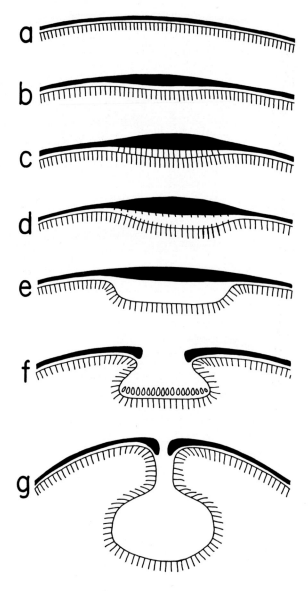

FIGURE 6. Diagrams showing stages in the development of type 1 conceptacles. A. No evidence of incipient conceptacle. B. Formation of cap by local thickening of cuticle. C. and D. Separation of cap and underlying tissue in consequence of the dissolution of the latter. E. Circumconceptacular growth resulting in formation of the walls of the conceptacular cavity and canal F. Sloughing of the cap and formation of reproductive initials on the floor of the conceptacle. G. Continued development resulting in a mature conceptacle with a cavity, canal, and pore. (Johansen, H. W., *Univ. Calif., Berkeley, Publ. Bot.*, 49, 1, 1969. With permission.)

Growth of those filaments originating adjacent to the disc form the ceiling of the chamber, whereas the more peripheral filaments grow inward and upward forming the bulk of the roof and the lining of the canal. In most conceptacles the canal and pore are directly above the center of the disc, but, in some taxa, circumconceptacular growth is asymmetrical and the pores are eccentric (e.g., *Arthrocardia duthiae, Chiharaea bod-*

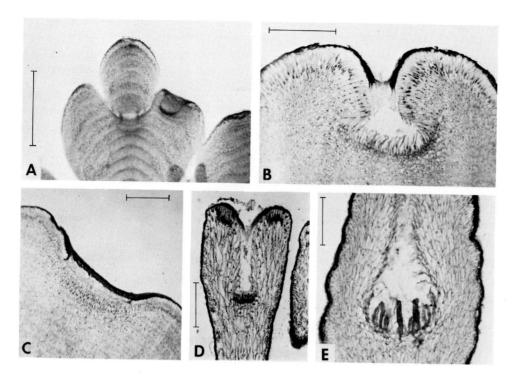

FIGURE 7. Immature tetrasporangial conceptacles of Type 1. A. *Cheilosporum*. The young conceptacle originated in medullary tissue, but not in line with the axis of branch. Scale = 0.5 mm. B. *Calliarthron tuberculosum*. Scale = 200 μm. C. Primordium in *Calliarthron*. Scale = 200 μm. D. Young tetrasporangia on the floor of conceptacle in *Jania*. Note the dense meristematic cells at top of intergeniculum. Scale = 150 μm. E. *Corallina*. Some premeiotic sporangia have elongated more than others. Scale = 100 μm. (Figs. 7A, B, and C, Johansen, H. W., *Univ. Calif., Berkeley, Publ. Bot.*, 49, 1, 1969; Figs. D and E, Johansen, H. W., *Br. Phycol. J.*, 5 (1), 79, 1970. With permission.)

egensis, *Calliarthron yessoense*, *Alatocladia modesta* and some forms of *Bossiella chiloensis*.

During development of a conceptacle, the part of the cap over the disc breaks away, probably due to the marked changes in configuration occurring beneath it. Usually the cuticular edge left where the cap broke free remains as a variously shaped rim around the conceptacular pore (Figure 6). The final shape of a conceptacle is variable and depends on the species.

As in other types of conceptacles, senescence may involve burial by vegetative tissues or the development of meristematic capabilities in cells lining the chamber (Figure 3). In the latter instance, large-celled tissue may completely fill old conceptacles.[223]

Type 2 (Figures 8, 9) — The second type of development takes place in conceptacles of the Amphiroideae, Mastophoroideae, Lithophylloideae, and Metagoniolithoideae. There are two indications of a conceptacle primordium: (1) secretion of cuticular material by perithallial or cortical meristematic cells (underneath the epithallus) and (2) the production of an increasing number of cells and the enlargement of certain cells with the result that a preconceptacular mound projects above the thallus surface. Within a mound the lowermost or first formed stratum of cells is longer than the several strata of short cells near the thallus surface (Figure 8). These long cells, the so-called *cavity cells*,[222] form a lens-shaped structure mostly responsible for protrusion of the mound.

Sporangial initials originate within a mound at the bottom level of the cavity cells when some of these begin to degenerate as the small initials grow into the vacated

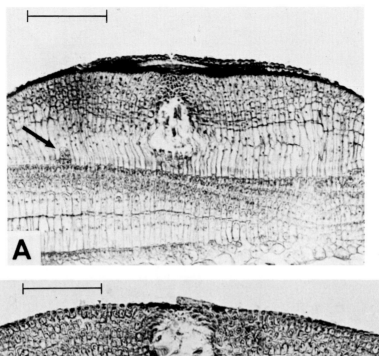
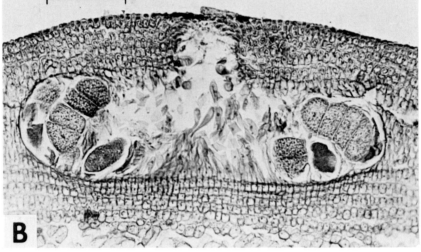

FIGURE 8. Tetrasporangial conceptacles in *Amphiroa ephedraea* (type II). A. Young conceptacle with sporangial initials (arrow) among degenerating cavity cells and developing pore. Scale = 150 μm. B. Mature conceptacle containing sporangia. Note remains of cavity cells forming central columella. Scale = 150 μm. (Johansen, H. W., *J. Phycol.*, 4(4), 319, 1968. With permission.)

space (Figure 8). The first initials are produced in a peripheral ring although in some taxa other initials later develop in the center of the mound. Each small initial divides transversely into a stalk cell and a premeiotic sporangium. The latter enlarge markedly and develop into tetrasporangia or bisporangia. As this development occurs, more cavity cells are crushed until even the peripherally located spores have access to the canal and pore at the center of the roof (Figure 8).

Details as to how a canal and pore form in a conceptacle roof which is intact prior to tetraspore formation are not known. Somehow there is degeneration of the centrally located roof cells and a loss of the mucilaginous cap overlying the roof. The filaments lining the canal become centripetally oriented. Probably the cavity cells in the center

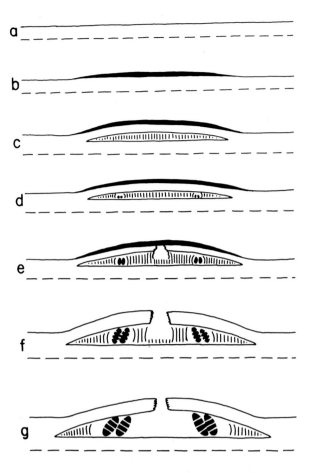

FIGURE 9. Diagram of stages of development of tetrasporangial conceptacle in *Amphiroa ephedraea*. A. Before initiation. B. Cap appears. C. Cavity cells elongate and a dome forms. D. Tetrasporangial initials appear. E. They enlarge and a pore begins to form. F. Continued development and degeneration of cavity cells. G. Mature conceptacle. (Johansen, H. W., *J. Phycol.*, 4(4), 319, 1968. With permission.)

of the conceptacle, usually the last to break down, play a role in canal formation. In some taxa these central cavity cells do not break down completely, instead they remain as a tuft of long cells called a *columella* (Figure 8).

Type 3 (Figures 10, 11) — Although minor differences in the development of tetrasporangial conceptacles occur among the genera of the Melobesioideae, a basic pattern holds for all. Recent studies have emphasized the developmental aspects of these kinds of conceptacles. Included among them are Adey and Johansen[23] on *Clathromorphum* and *Mesophyllum*, Lebednik[262] on *Clathromorphum*, Woelkerling,[446] on *Mastophoropsis*, and Townsend[435a] on *Synarthrophyton*.

Early in conceptacle development, a group of perithallial meristematic cells (or cells below the meristem in *Phymatolithon* and *Leptophytum*) begin elongating to form a stratum of cavity cells longer than the surrounding cells. During this growth, divisions occur above the cavity cells and the roof of the future conceptacle begins to appear. Eventually a lens-like layer of cavity cells in which the central cells are longest is covered by several layers of small roof cells (Figure 10).

A fraction of the meristematic cells is destined to become sporangial initials. These

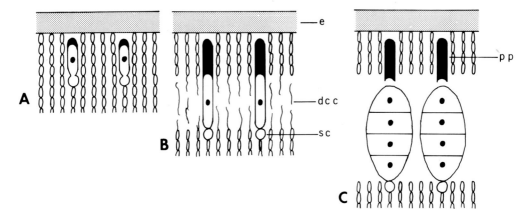

FIGURE 10. Diagrams of stages in the development of tetraporangia in *Clathromorphum*. Epithallial tissue (e) is shaded. A. initial cells are dividing into stalk cells and young sporangia in which apices are thickening. B. A later stage showing stalk cells (sc) and sporangia with thick apices among cells of the conceptacle roof. Cavity cells are degenerating (dcc). C. Part of mature conceptacle with pore plugs (pp) separated from the sporangia.

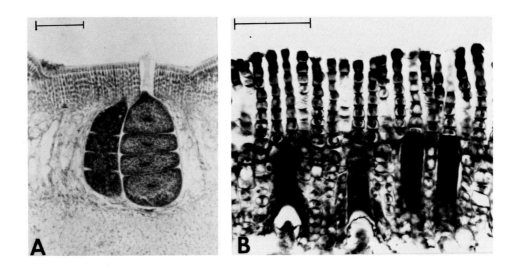

FIGURE 11. A. Tetrasporangia in *Mesophyllum conchatum*. Scale = 100 μm. B. Roof of tetrasporangial conceptacle in *Clathromorphum nereostratum* showing epithallial cells above the pore plugs. Scale = 50 μm. (Fig. A, Adey, W. H. and Johansen, H. W., *Phycologia*, 11(2), 159, 1972; Fig. B, Lebednik, P. A., *Syesis*, 9(x), 59, 1977. With permission.)

do not participate in forming cavity cells or roof cells. Instead, these initials elongate without dividing during which time densely staining mucilaginous apices appear (Figure 11). Sometime during this development, initials divide to become cell duos, each consisting of a young sporangium subtended by a small stalk cell (Figure 10). The sporangia are at the same level as the cavity cells and are surrounded by the cells or their remains. Continued enlargement of the sporangia maintains the densely staining mucilaginous apices among the small roof cells.

Meioses, cytokineses, and enlargement of the sporangia occur concomitant with a crushing and degeneration of the cavity cells within the dome. Discharge of a sporangium is preceded by expulsion of the thick pore plug which, in effect, leaves a canal through which exit is made. The forces involved in expelling sporangia through the relatively small canals are not understood.

To summarize, it is possible to list an approximate chronology of the events leading to conceptacle formation in the Melobesioideae, recognizing of course, that there is overlapping of activity.

1. The formation of a stratum of cavity cells, called an elongation layer by Lebednik[262]
2. Divisions at the apices of these cells to form small roof cells
3. The enlargement of sporangial initials located among the cavity cells
4. The division of each of these cells into a premeiotic sporangium and subtending stalk cell
5. The thickening of sporangial apices into mucilaginous pore plugs
6. The continued elongation of the sporangia so that the mucilaginous apices extend into and among the roof cells
7. The degeneration of the long cavity cells among the developing sporangia
8. The sloughing off of epithallia above the conceptacles
9. Meiosis in the sporangia
10. Cytokineses, the divisions of young sporangia into tetrasporangia (or bisporangia)
11. Spore or sporangium discharge preceded by expulsion of pore plugs or the sloughing off of conceptacle roofs
12. Conceptacle senescence, often involving burial

Most authors claim that each distinctive pore plug in the Melobesioideae originates as part of the sporangial apex.[23] Yet, Lee[267] found that in *Melobesia mediocris* the plugs were differentiated from one or two cells distal to each tetrasporangium. He called them "pore cells." On the basis of very little evidence, it is assumed that the sporangia are extruded through the pores after they are unplugged. In *Synarthrophyton patena,* however, the entire conceptacle roof falls away prior to spore release.[435a]

Type 4 (Figure 5, Chapter 1) — These conceptacles occur only in asexual plants of *Sporolithon*. They are usually produced in groups, and due to the fact that only a single sporangium is housed within a chamber, the groupings have been called "nemathecia" rather than conceptacles.[449] As far as it is known, development is the same as in Type 3, except that the vegetative filaments among the sporangia do not break down to include the reproductive cells in a common chamber.

TETRASPORANGIA AND BISPORANGIA

The developmental stages in the transformation of initial cells to fully-formed tetrasporangia or bisporangia are (1) division of sporangial initials into young sporangia and subtending stalk cells, (2) enlargement of the sporangia and preparation of the nuclei for meiosis, (3) meiosis and the alignment of the resulting nuclei on the cell axes (meiosis does not occur in some bisporangia), and (4) further enlargement and transverse cytokineses, resulting in two or four spores per sporangium.

As alluded to earlier, the reproductive initials are variously located in different taxonomic groups. In the Corallinoideae they form a uniform layer in the floor of the conceptacle primordium, under the cap. In the Amphiroideae the initials form a ring and sometimes a central patch underneath a dome of tissue (Figure 8). In the Melobesioideae they are located in a cluster among vegetative filaments. Despite these positional differences, early development is possibly similar in all taxa except *Sporolithon* where cruciately divided tetrasporangia have been reported.[449]

The divisions that result in two-celled consortiums of tetrasporangia (or bisporangia) and subtending stalk cells occur when the conceptacles are still very young. The stalk cells remain small and do not appear to change further. The young sporangia do have the capacity to enlarge markedly, although a significant number do not do so. Possibly there is room for a limited number of sporangia in each conceptacle, and only the precocious cells mature. This seems to be especially true in *Jania* and related genera where there are fewer than ten large tetrasporangia per conceptacle even though numerous underdeveloped ones are present.

The enlargement of young sporangia about 30 μm in diameter and 15 to 25 μm long, proceeds until the dimensions prior to meiosis are about 15 to 25 μm and 70 to 90 μm (e.g., *Calliarthron tuberculosum*).[223] By this time, the cytoplasm is dense and the nucleus large.

Meiosis is apparently a relatively rapid event for rarely are the different stages seen. Over 80 years ago, Davis[109] described and illustrated meiosis in *Corallina elongata* (as *C. officinalis* var. *mediterranea*). More recently Magne[293] examined the phenomenon in *Corallina officinalis*, confirming Davis's work as well as that of Yamanouchi[453] and Westbrook.[445] As prophase I proceeded, Magne was hampered by unusually feeble staining characteristics. As far as it is known, meiosis is similar to that usually described for higher plants. It should be noted that immediately prior to the first division the nuclei enlarge markedly, becoming about 15 μm in diameter. Some stages in meiosis have also been illustrated. These include: *Jania rubens*[421] (as *Corallina rubens*) and *Fosliella farinosa*[35] (as *Melobesia farinosa*.)

A close look at post-meiotic tetrasporangial nuclei was taken by Peel et al.[351] who examined, with the electron microscope, the sporangia of *Corallina officinalis* before cytokinesis had occurred. Tetrasporangia are cells undergoing marked changes both as regards nuclear phenomena associated with meiosis and the concomitant large increases in size. Peel et al.[351] found precisely aggregated endoplasmic reticulum surrounding the nuclei (nuclear endoplasmic reticulum), an electron-dense substance in contact with the nuclear surface, and extending among the nuclear endoplasmic reticulum, highly porous nuclear envelopes, intranuclear membrane systems, vacuolar regions in nucleoli, and large numbers of ribosomes in the immediate vicinity of the nuclei. All these evidences suggest hyperactive nuclei and close nuclear/cytoplasmic interaction involving a surge of ribosome synthesis associated with the remarkable changes that occur during tetrasporogenesis.

Once the nuclei are in line along the tetrasporangial axis, cytokineses commences. Infurrowing of an inner wall layer begins simultaneously at three places equidistant from one another. Cytokineses proceeds relatively slowly, as shown by the frequency with which the stages are found. Completion of this process results in four separate cells in which pit-connections are not evident. An outer wall layer still surrounds the four spores. This wall may remain intact even until and after discharge.

Fully-formed tetrasporangia and bisporangia are large, but their sizes vary somewhat among the taxa. In the Lithophylloideae and Amphiroideae, they are relatively small: 40 to 60 μm by 85 to 110 μm. In the Corallinoideae, they are larger, and in *Jania pusilla* they are 130 to 230 μm by 200 to 350 μm.[147] Bisporangia are generally as large as tetrasporangia in the same species, therefore the bispores are larger than the tetraspores.

MALE CONCEPTACLES

In many taxa male conceptacles may be externally recognized as such under low magnification. Partly due to the absence of a chamber filled with sporangia, these

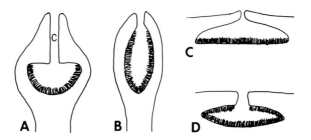

FIGURE 12. Outlines of male conceptacles. A. *Corallina*. Note the long conceptacular canal (c). B. *Jania*. C. *Amphiroa*. D. *Clathromorphum*.

conceptacles are usually uniquely shaped (Figures 12, 13). In some articulated coralline algae, conceptacle development is such that a long canal passes through a beak-like roof. This is very pronounced in *Arthrocardia* and less so in *Corallina* (Figure 12). At least in some coralline algae, male conceptacles (and female conceptacles also) are more crowded than they are in tetrasporangial or bisporangial plants.

With the exception of the Melobesioideae, male conceptacles in the coralline algae exhibit the same basic plan of development. A group of meristematic cells becomes differentiated into spermatangial initials which form a coherent layer over the floor and sometimes the walls of the developing chamber. Each initial cuts off one to three spermatangial mother cells that then form a dense, deeply staining layer constituting the fertile area of the conceptacle. These spermatangial mother cells cut off spermatangia, as we shall see later. During development of the male cells vegetative filaments surrounding the fertile disc grow centripetally so as to form a uniporate conceptacle similar to tetrasporangial conceptacle type 1. The cells subtending the spermatangial mother cells may be indistinguishable from other vegetative cells or they may be recognized as different by their denser cytoplasm.

Considerable variety in conceptacle shape occurs in the Corallinaceae and a sampling of these shapes is shown in profiles in Figure 12. Note the extents and placements of fertile areas in *Amphiroa* where only the floor of the chamber is covered, and in *Corallina* where the fertile surface extends up the walls of the chamber. These characteristics serve to emphasize the placement of these genera in separate subfamilies. Furthermore, note the difference between the male conceptacles in *Corallina* and *Jania*. This difference is one of the characteristics distinguishing the two tribes in the Corallinoideae.[236]

Now our attention turns to the diverse types of male conceptacles in the Melobesioideae. In a recent paper, Lebednik[264,265] summarized the current status of knowledge on male conceptacle development in this subfamily. Unlike the other subfamilies, several patterns of development have evolved and prove to be taxonomically useful at the generic level. Five patterns of development are exemplified by the genera: *Lithothamnium, Phymatolithon, Leptophytum, Mesophyllum,* and *Clathromorphum*.

In all genera of the Melobesioideae, spermatangial mother cells occur on the roof as well as on the floor of the mature conceptacle. Lebednik[264,265] pointed out that four significant features should be considered in evaluating conceptacle development in this subfamily.

1. The mode of origin of spermatangial initials. As shown in Figure 14, the spermatangial initials form directly from meristematic cells in *Lithothamnium, Mesophyllum,* and *Clathromorphum*. These cells are formed adventitiously from perithallial cells below the meristematic cells in *Phymatolithon* and *Leptophy-*

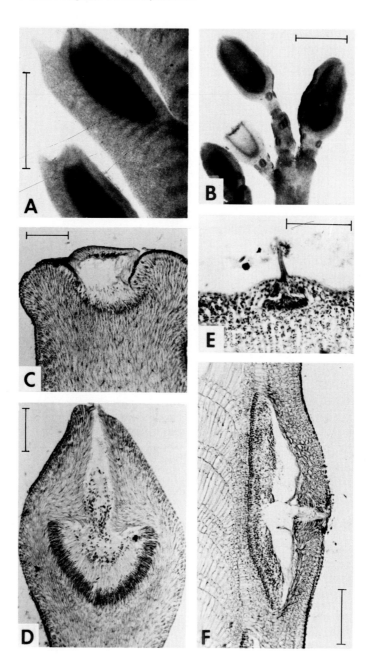

FIGURE 13. Male conceptacles. A. *Cheilosporum*. In this photograph and B. The specimen was decalcified and photographed whole. The chamber is dark. Scale = 1 mm. B. *Jania*. Note the young conceptacle. Scale = 400 μm. C. Young conceptacle of *Corallina officinalis*. Scale = 100 μm. D. Mature conceptacle of *Corallina officinalis*. Scale = 100 μm. E. *Porolithon boergesenii*. Note acellular projection, or spout. Scale = 100 μm. Courtesy T. Masaki. F. *Amphiroa ephedraea*. Scale = 100 μm. (Fig. A, Johansen, H. W., *J. S. Afr. Bot.*, 43(3), 163, 1977; Figs. B, C, D, and E, Johansen, H. W., *Br. Phycol. J.*, 5(1), 79, 1970; Fig. F, Johansen, H. W., *J. Phycol.*, 4(4), 319, 1968. With permission.)

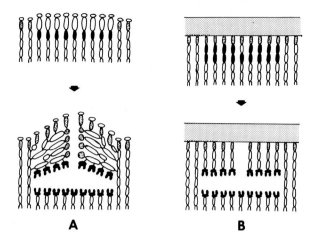

FIGURE 14. Diagrams of young and old stages in the development of male conceptacles in two genera of the Melobesioideae. Meristematic cells are shaded and reproductive initials, spermatangial mother cells and spermatangia are blackened. A. *Mesophyllum*. A special layer of cells occurs above the reproductive initials. The conceptacle roof forms by the growth of filaments from around the fertile area. B. *Clathromorphum*. The epithallus is shaded. The conceptacle roof develops from filaments produced by meristematic cells above the reproductive initials. Adapted from Lebednik.[264-265]

tum. In these last two genera, the vegetative tissue above the initials is sloughed off during development. In this regard, development resembles that in the tetrasporangial and female conceptacles (where known) in these two genera.

2. The presence of a protective layer over the spermatangial initials early in their development. A protective layer of ephemeral cells forms above the initials in *Mesophyllum* (and *Leptophytum*?) as described for the first time by Lebednik.[264,265]
3. The mode of formation of the conceptacle roof. In *Clathromorphum* and *Melobesia* the conceptacle roof is probably formed by continued growth of the filaments that form the spermatangial initials (Figure 14). In the other genera meristematic activity terminates when the filaments give rise to spermatangial initials, and then the roof is formed by centripetal growth of the vegetative filaments surrounding the fertile disc.
4. The arrangement of the mature spermatangial mother cells. In three genera, *Lithothamnium, Phymatolithon,* and *Synarthrophyton*,[465] the spermatangial mother cells divide so as to form dendroid branching systems (Figure 15) rather than simple strata of spermatangial mother cells. *Leptophytum* is intermediate in that some of the filament systems are dendroid whereas others are simple. Lebednik[264,265] pointed out that developmental studies of male conceptacles are needed, especially in generic type species in the Melobesioideae, to elucidate possible evolutionary trends in this subfamily. It is now prudent to examine closely the development of *Mesophyllum* and *Clathromorphum,* two genera in which male conceptacles were studied by Lebednik.[262,264,265]

The development of male conceptacles in *Mesophyllum aleuticum* and *Mesophyllum conchatum* was followed by Lebednik[264,265] and is diagrammed in Figure 14. In the

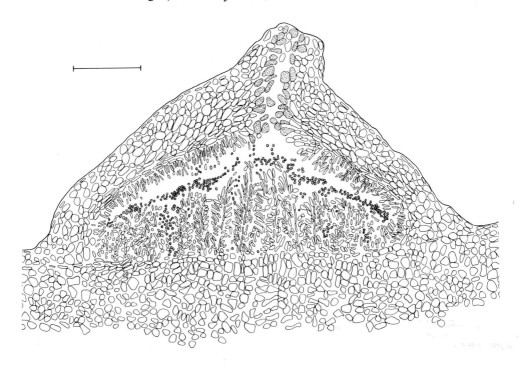

FIGURE 15. Male conceptacle of *Lithothamnium pacificum* with dendroid spermatangial clusters. Scale = 50 μm. Courtesy T. Masaki.

earliest recognizable stage, the epithallial cells separate from an underlying group of meristematic cells as the apical walls of the meristematic cells gelatinize. These meristematic cells remain viable, and each divides so as to form a lower and an upper cell. The lower cells constitute a stratum of densely cytoplasmic spermatangial initial cells with large nuclei. The upper cells are a protective layer of long, highly vacuolate cells that produce copious amounts of mucilage on the surface of the thallus. This cell layer and the secreted mucilage are the cap of the young conceptacle. As development proceeds, the protective cells degenerate and disappear. Meanwhile, a roof is forming by the centripetal growth of surrounding vegetative filaments. The later stages of development include the formation of spermatangial mother cells from the initials on the floor of the chamber as well as from filaments lining the roof of the chamber. At maturity a conceptacle chamber has columnar spermatangial mother cells producing spermatia on the floor and roof.

The following description of the unique male conceptacles in *Clathromorphum* is mostly from Lebednik's studies[262] of *C. nereostratum* and *C. reclinatum*. The first evidence of a conceptacle is the formation of spermatangial initial cells by basipetal divisions of several perithallial meristematic cells near the margin of a thallus. Each meristematic cell divides twice, first forming a spermatangial initial cell destined for the floor of the chamber and then cutting off another that will participate in a fertile layer lining the roof of the chamber (Figure 14). These divisions continue until a considerable number of floor and roof spermatangial initial cells have been cut off. At this time, pore initiation is indicated by three observations:[262] (1) a few centrally-located meristematic cells fail to cut off roof initials, (2) there is a loss of epithallial filaments in this central area, and (3) there is some localized intensity of staining.

Two to five spermatangial mother cells are formed by oblique divisions of each spermatangial initial cell with the remaining proximal cell serving as a basal cell. At about this time, the fertile cell layers split apart and a conceptacle becomes recognizable as such. Some of the perithallial meristematic cells at the center of the conceptacle have, by this time, lost their overlying epithallial cells and have begun proliferating filaments which secrete mucilage into the developing pore. Eventually these meristematic cells degenerate. The production of sterile cells by the perithallial meristem results in cells being added to the conceptacle roof (for a total of six to seven cell divisions after the formation of roof initials in *C. nereostratum*). Now the conceptacle may be considered to be fully mature (Figure 14). Two to five spermatangial mother cells per basal cell occur on the floor and roof, but not on the sides, of the chamber. Spermatia are cut off obliquely from the tips of the mother cells in a uniseriate manner. The crushing of cells immediately above and below the chamber and the elongation of perithallial cells around the conceptacle results in an elliptical chamber. The chamber diameter varies from 180 to 400 μm and the height is from 80 to 220 μm. Buried senescent conceptacles were also observed by Lebednik.[262]

These developmental features probably hold for male conceptacles in all species of *Clathromorphum*. In addition to Lebednik's[262] studies, observations have been made on *C. circumscriptum*,[4-5,306-307] *C. reclinatum* and *C. parcum*.[23]

SPERMATANGIA

There are three types of cells in a fertile layer on the inner surface of a male conceptacle: basal cells, spermatangial mother cells, and spermatangia (Figure 16). The enormous numbers of spermatangia are deciduous and function as dispersal units. They may be dispersed in slime[330] and pass through the water in masses, rather than individually. Probably they do not travel far, for female plants and male plants usually occur together. Whether spermatangia themselves fuse with trichogynes, or they produce spermatia which do so, is an open question. In *Jania rubens* Suneson[422] observed spermatangial protoplasts being released and suggested the possibility that they are functional spermatia (Figure 17).

Peel and Duckett[350] described ultrastructural features and cytochemical studies of male reproductive systems in *Corallina officinalis* from Great Britain. Basal cells are modified vegetative cells with their most distinctive feature being a curved rod-like structure composed of parallel microtubules appearing to extend the length of each cell, often near the nucleus. Spermatangial mother cells are shorter than basal cells and are notable in that the plastids degenerate and the cell walls contain a distinctive fibrillar layer. These cells give rise to the elongate, closely-packed spermatangia which eventually become detached. During development, the spermatangial protoplasts shorten into the distal parts of the cell wall with the empty proximal parts of the wall forming an elongate "tail" (Figure 17). The nuclear material condenses into two densely staining bands that have also been recorded in studies with light microscopes (*Corallina officinalis* and *Jania rubens*).[421] Suneson first suggested[421] that they were nuclei undergoing mitosis, but later, in light microscopical studies of *Metamastophora lamourouxii*,[426] he felt that the two nuclear groups might be fixation artifacts. Peel and Duckett[350] believe that the distinctive intranuclear bodies in the spermatangia (the Körnchen of Grubb, 1925) are groups of chromatids or chromosomes in an anaphase stage of arrested mitotic division. Stages later than anaphase have not been seen.

Plastids seemingly degenerate during the formation of spermatangia until these cells lack them completely or almost so. Starch grains are plentiful, the raw material probably being obtained from outside the cells.

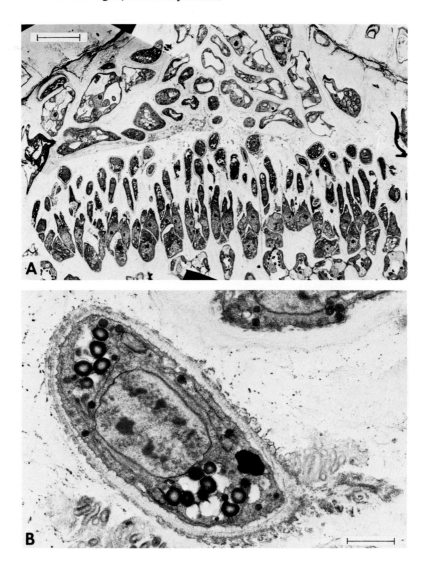

FIGURE 16. Electron micrographs of spermatangia and related structures in *Lithothrix aspergillum*. A. Spermatangia being produced from spermatangial mother cells Scale = 15 μm. B. Spermatangium showing the condensed chromatin, endoplasmic reticulum around the nucleus, microtubules between the nuclear envelope and the endoplasmic reticulum, starch grains but no plastids, and a thick cell wall with plasmodesmata projecting through it. Scale = 1 μm. Both courtesy M. A. Borowitzka and M. Vesk.

CARPOGONIA

In all female conceptacles that have been studied, development involves the formation of a uniform layer of carpogonial initials by all of the meristematic cells of the disc. A conceptacle roof is always produced by circumconceptacular filaments which continue to grow in length while the disc cells become wholly involved in producing reproductive cells (Figure 18).

Each meristematic cell of a disc filament divides to form one to three carpogonial filament initials (only some of which will become carpogonial filaments) and a sub-

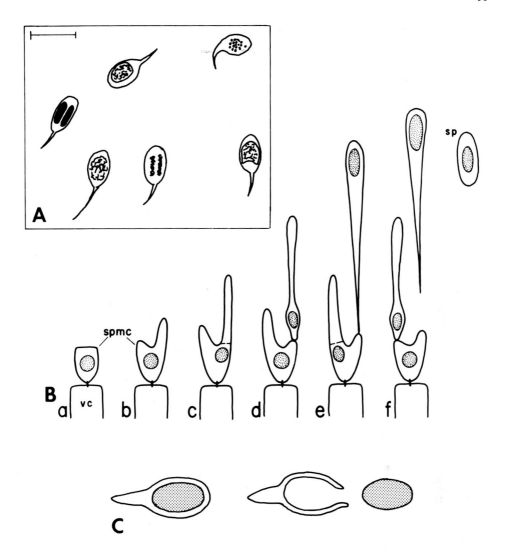

FIGURE 17. Spermatangia. A. Drawings showing various nuclear conditions as seen with the light microscope. Scale = 5 μm. B. Diagrams of stages in the development of spermatangia as seen in Calliarthron. Spmc = spermatangial mother cells; Sp = spermatangia. C. Spermatangium releasing spermatium (shaded) as observed in *Jania rubens* by Suneson.[422] (Fig. 17B, Johansen, H. W., *Univ. Calif., Berkeley, Publ. Bot.*, 49, 1, 1969. With permission.)

tending supporting cell. These divisions begin at the center of the disc with succeeding divisions proceeding centrifugally. Numerous taxa have been shown to have fully developed carpogonial filaments in the disc centers and undeveloped ones in the peripheries (reported in 20 genera).[263] The peripheral filaments lack trichogynes, usually lack hypogynous cells, and tend to branch into two or three cells even though in the center of the disc supporting cells may only produce single filaments (Figure 19).[223] These peripheral filaments are usually considered to be reduced carpogonial filaments, and recently Lebednik[263] has shown that their subtending supporting cells are auxiliary cells in some taxa. This idea will be discussed later. Fully developed carpogonial filaments in disc peripheries have been reported in a few taxa.

Suneson,[421] Johansen,[223,230] and Lebednik[263] have described in detail the development of carpogonial filaments. Where only single initials are formed per supporting

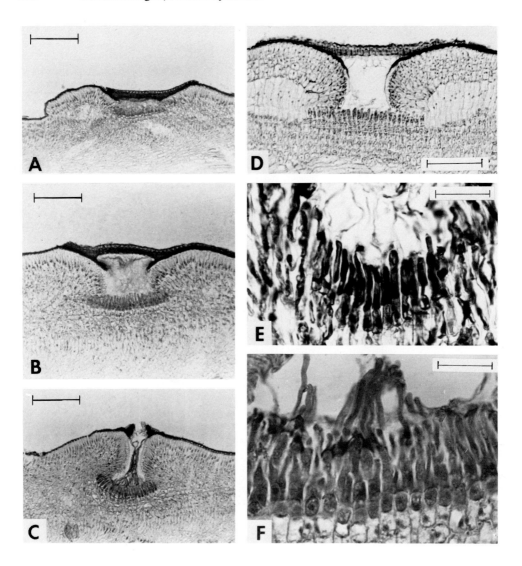

FIGURE 18. Female conceptacles. A - C. *Bossiella californica* ssp. *schmitti,* from young to mature. Scales = 100 μm. D. Young conceptacle of *Amphiroa ephedraea* with cap still intact. Scale = 100 μm. E. *Cheilosporum cultratum* with young procarps. Scale = 30 μm. F. Procarps of *Calliarthron tuberculosum.* The deeply stained lowermost cells are supporting cells. Scale = 30 μm. (Figs. A, B, and C, Johansen, H. W., J. Phycol., 9(2), 141, 1973; Fig. D, Johansen, H. W., J. Phycol., 4(4), 319, 1968. With permission.)

cell, the plane of division is transverse, such as in *Calliarthron tuberculosum*[22,3] and *Mesophyllum.*[263] Where two or three are produced per supporting cell, the divisions are oblique, such as in *Clathromorphum.*[263] Carpogonial filament initials enlarge slightly, the apices elongate to become trichogynes, and transverse divisions occur, resulting in units containing an ultimate carpogonium and a penultimate hypogynous cell (Figure 20). A fully formed carpogonial filament consists of a carpogonium, with a swollen, nucleus-containing base and a long trichogyne, and a small hypogynous cell. A filament may be located singly on a supporting cell or with one or two other filaments that are usually just single sterile cells lacking trichogynes. At the time trichogynes are receptive to contact by spermatia, some of them project through the conceptacular pores.

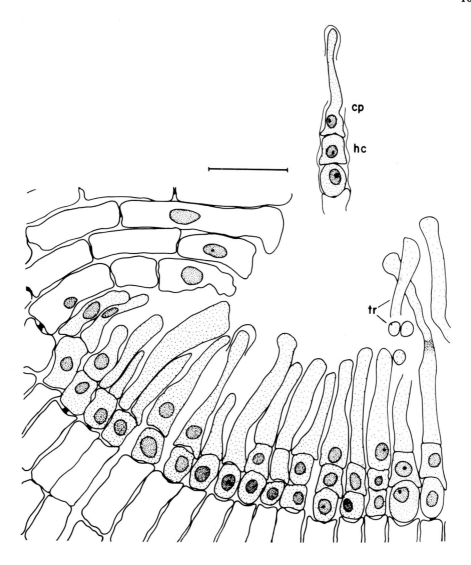

FIGURE 19. Carpogonial filaments and supporting cells in *Calliarthron tuberculosum*. Arrow indicates center of floor of conceptacle. Some filaments at the margins lack trichogynes and hypogynous cells. A complete carpogonial filament on a supporting cell is shown above: cp = carpogonium; hc = hypogynous cell. Scale = 20 μm. (Johansen, H. W., *Univ. Calif., Berkeley, Publ. Bot.*, 49, 1, 1969. With permission.)

CARPOSPOROPHYTES

The post-fertilization development of a parasitic carposporophyte with diploid cells occurs within female conceptacles. This is a complex process that involves cellular fusions and the production of gonimoblast filaments (= carposporangial filaments) containing carposporangia. The growth of gonimoblast filaments occurs concomitant with enlargement of the female conceptacle.

The fusion of spermatium and trichogyne has not been reported for coralline algae, nor has the migration of a male nucleus to a carpogonial nucleus. Probably one or several zygotic nuclei are produced in each female conceptacle when fertilization occurs. One or more of these nuclei or their derivatives migrates to the subtending supporting cell, which may be considered the auxiliary cell. The arrival of a nucleus prob-

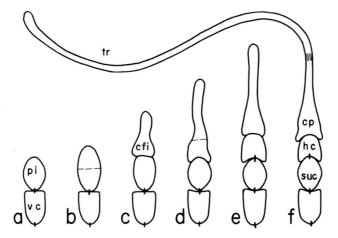

FIGURE 20. Stages in the development of a procarp, such as in *Calliarthron tuberculosum*. pi = procarp initial; vc = vegetative cell; cfi = carpognial filament initial; cp = carpogonium; hc = hypogynous cell; suc = supporting cell. (Johansen, H. W., *Univ. Calif., Berkeley, Publ. Bot.*, 49, 1, 1969. With permission.)

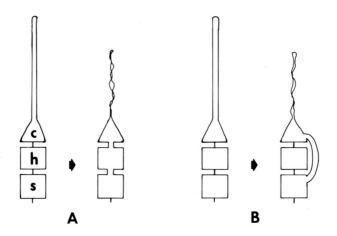

FIGURE 21. Two possible routes by which zygotic nuclei pass from a carpogonium to the supporting cell below. A. Cellular fusions. B. Transfer tube.

ably triggers the fusions of the supporting cell and neighboring cells. There are at least three routes by which the zygotic nuclei, or their derivatives, pass from the carpogonial bases to the supporting cells below (Figure 21).

1. Carpogonia which have probably been fertilized have been observed to have fused with their subtending hypogynous cells which, in turn, are attached to the fusion cell by a narrow canal. This suggests that dipoloid nuclei pass directly downward through these fused cells. This appears to be the pathway in *Calliarthron tuberculosum*,[223] *Bossiella californica* ssp. *schmittii*,[230] *Amphiroa ephedraea*,[228] *Corallina elongata*,[453] (as *C. officinalis* var. *mediterranea*), *Phymatolithon rugulosum*,[3] *Lithothamnium glaciale*,[20] and *Metamastophora flabellata*.[465]
2. Thin, unseptated *transfer tubes* extending downward from carpogonial bases to subtending supporting cells have been reported and illustrated by Suneson,[421]

Ganesan,[171] and recently by Lebednik.[263] These transfer tubes have been proposed to be the means by which diploid nuclei are carried directly to cells which initiate the fusion cell; they bypass the intervening hypogynous cells. The existence and role of transfer tubes is still based on insecure evidence; they are extremely delicate and apparently exist for only a short time in fertilized conceptacles. Lebednik[263] suggested that they may be more prevalent than formerly realized, at least in the Melobesioideae. On the other hand, Woelkerling[465] questioned the existence of transfer tubes, pointing out their delicate nature and the insecure evidence for them.

3. An upward directed process from the auxiliary cell to the carpogonium has been reported for *Choreonema thuretii* by Minder[324] and Suneson.[421]

For many years post-fertilization development in coralline algae has been shown to involve the formation of a coenocytic fusion cell from which the gonimoblast filaments arise. A fusion cell is probably initiated by a supporting cell into which a diploid nucleus has passed. Three situations exist:

1. In most genera belonging to the Corallinoideae, Amphiroideae, Mastophoroideae, and Lithophylloideae, a single, relatively coherent fusion cell covers nearly the entire fertile area in the base of the conceptacular chamber. This cell is an amalgamation product of supporting cells, with the perithallial or cortical cells below remaining uninvolved.
2. In most genera in the Melobesioideae, a discontinuous, irregularly-shaped fusion cell or several small, ill-defined fusion cells have been reported. Lebednik[263] showed that fusion cells in some of these taxa are limited to the center of the disc in the area where carpogonial filaments were located prior to fertilization. In some species of *Clathromorphum* and *Mesophyllum* other cells below the supporting cells also become incorporated in the fusion cells.[263,421] These may appear a little different from other perithallial cells, or they may be densely cytoplasmic.
3. Finally, in a few taxa of the Melobesioideae, a true fusion cell may be lacking, such as in *Melobesia mediocris*[267] and *Phymatolithon lenormandii*.[423] Further study may reveal that several small fusion cells do exist.

Passage of diploid nuclei into supporting cells near the center of a disc probably stimulates these cells to begin fusing with neighboring supporting cells. Sections through fusion cells reveal numerous nuclei, and it has been suggested[223] that both large diploid nuclei as well as small haploid nuclei are included in these cells. This would be expected.

The function of carposporophytes is the production of carpospores. These are produced in gonomoblast filaments, each of which in coralline algae consists of a graded series of cells, the most proximal of which is the smallest and the most distal is the large, red carposporangium. Development involves the formation of plastids and starch grains as well as a large increase in size. Borowitzka[50] described plastid and starch grain development in gonimoblast filaments of *Lithothrix aspergillum* (see Chapter 2).

Following establishment of fusion cells, there appear to be two ways in which gonimoblast filaments may be produced:

1. Gonimoblast filaments grow directly from fusion cells and do not fuse with other cells following their initiation. These filaments may grow only from the margins, as in the Lithophylloideae and scattered taxa in the other subfamilies, or they may grow from the margins as well as the upper surfaces, as is especially evident

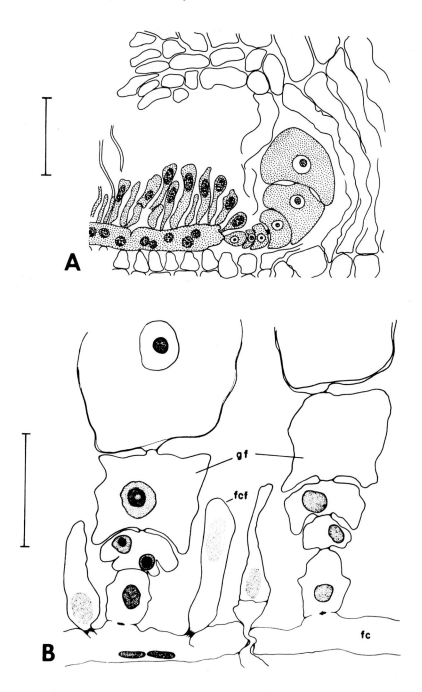

FIGURE 22. Fusion cells and gonimoblast filaments. A. *Amphiroa ephedraea*, showing gonimoblast filaments arising from margins of a fusion cell. Scale = 30 μm. B. In *Calliarthron tuberculosum* the gonimoblast filaments (gf) arise from the surface of the fusion cell (fc). Note the remains of two carpogonial filaments (fcf) in which the hypogynous cells have disappeared and intimate connections have become established with the fusion cell. Scale = 30 μm. (Fig. A, Johansen, H. W., *J. Phycol.*, 4(4), 319, 1968; Fig. B, Johansen, H. W., *Univ. Calif., Berkeley, Publ. Bot.*, 49, 1, 1969. With permission.)

in some corallinoidean genera such as *Arthrocardia*, *Calliarthron*, and *Bossiella* (Figures 22, 23).

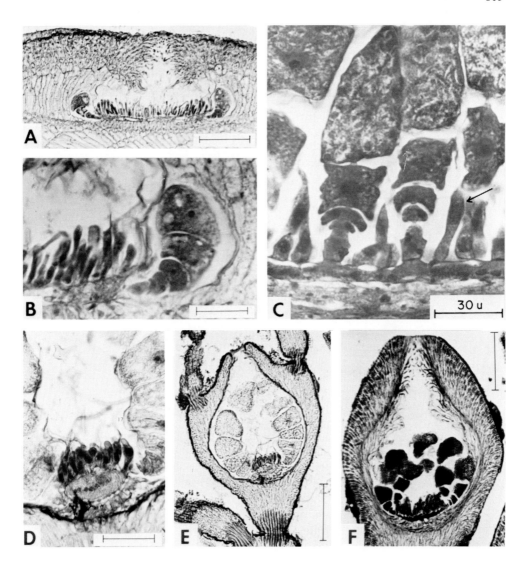

FIGURE 23. Fusion cells and gonimoblast filaments. A. *Amphiroa ephedraea* in which gonimoblast filaments grow from the margins of the fusion cell. Scale = 200 μm. B. Enlarged view of one filament. Scale = 25 μm. C. *Calliarthron tuberculosum* in which gonimoblast filaments arise from the upper surface of the fusion cell. Arrow points to remains of carpogonial filament. Scale = 30 μm. D. Thick fusion cell in *Haliptilon subulatum*. The gonimoblast filaments are marginal and the remains of carpogonial filaments stud the upper surface of the fusion cell. Scale = 50 μm. E. Another view of D. Scale = 150 μm. F. *Corallina officinalis*, with a broad, thin fusion cell and marginal gonimoblast filaments. Scale = 150 μm. (Figs. A and B, Johansen, H. W., *J. Phycol.*, 4(4), 319, 1968; Fig. C, Johansen, H. W., *Univ. Calif., Berkeley, Publ. Bot.*, 49, 1, 1969; Figs. D, E, and F, Johansen, H. W., *Br. Phycol. J.*, 5(1), 79, 1970. With permission.)

2. Lebednik's studies[263] of *Clathromorphum, Mesophyllum,* and *Melobesia* showed for the first time that carposporangium producing filaments fuse with cells as they grow (Figures 24, 25). He described these filaments as establishing open connections with peripherally located supporting cells as they grow radially from small, centrally located fusion cells (Figure 25). Lebednik interpreted these supporting cells as auxiliary cells, and suggested that these taxa were not procarpial. Papenfuss[346] defined procarps as filament systems containing both carpogonia

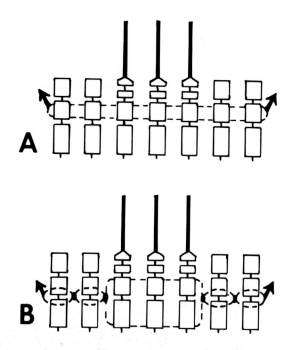

FIGURE 24. Diagrams showing two ways in which fusion cells (dashed lines) are formed. Arrows show where gonimoblast filaments are produced. A. Conventional style, in which supporting cells fuse to form a broad disc-like cell. B. The type shown by Lebednik[263] to occur in at least some of the Melobesioideae. The fusion cell incorporates basal cells as well as supporting cells and produces gonimoblast filaments (connecting filaments) that fuse with other cells of the fertile disc before producing carposporangia.

and auxiliary cells. Lebednik[263] called the radially growing filaments *connecting filaments*. When they reach the periphery of the disc, they cut off carposporangia.

Woelkerling[466] discussed Lebednik's interpretation and pointed out that several non-corallinaceous red algae produce gonimoblast filaments that also fuse with various cells as they grow. He therefore sees no reason for considering the connecting filaments of Lebednik as anything other than gonimoblast filaments that fuse with other cells prior to producing carposporangia. Townsend[435a] in a study of *Synarthrophyton patena*, another melobesioidean taxon, also reported gonimoblast filaments as fusing with cells as they grow. Townsend also did not accept Lebednik's connecting filament idea. Other taxa in which gonimoblast filaments may fuse with cells are *Leptophytum*,[8] *Lithothamnium*,[20,303] and *Kvaleya*.[28]

LIFE HISTORIES

Although the evidence showing that carpospores germinate into tetrasporophytes and tetraspores into gametophytes is mostly absent, coralline algae clearly have a *Polysiphonia* type of life history.[136] Von Stosch[420] came the closest to demonstrating this when he cultured tetraspores of *Corallina officinalis* from which gametophytes developed. Fertilization of these female plants resulted in carpospores which grew into

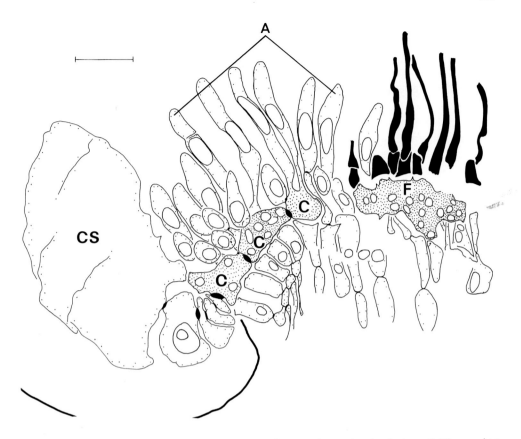

FIGURE 25. Part of a carposporophyte of *Mesophyllum conchatum* showing fusion cell (F), a gonimoblast filament, or connecting filament (C), and a carposporangium (CS). A = remains of auxiliary cell filaments at periphery of conceptacle. The remains of carpogonial filaments are black. Scale = 25 μm. (Lebednik, P. A., *Phycologia*, 16(4), 379, 1977. With permission.

sterile plants that probably were tetrasporophytes lacking reproductive cells. The results of von Stosch's[420] culture work with *Jania rubens* were also interesting. After he modified the concentrations of iron, nitrogen, and phosphorus, he was able to obtain fertile gametophytes from tetraspores. At one point he obtained tetrasporophytes which also bore male conceptacles. These plants were diploid and meiosis resulted in haploid tetraspores which germinated and grew into second generation gametophytes which produced carpospores. These carpospores produced some normal tetrasporophytes and some which again also bore male conceptacles. Apparently at the previous meiosis there had been a segregation into some tetraspores whose progeny were normal tetrasporophytes and others where the tendency for maleness was retained. It is indeed surprising that in the only study of this type in coralline algae these results should have been obtained.

Based on the numbers of plants collected, far more coralline tetrasporophytes occur in the seas than do gametophytes.[421] The reason for this disparity is not known, for in culture experiments tetraspores usually germinate very readily. Part of the answer may lie in the seasons when the plants are collected. Chihara[100] showed that in Japan plants were mostly fertile (probably tetrasporangia) in the autumn an winter except in the Lithophylloideae and Amphiroideae which reproduce in the summer. Also, in Japan, Adey et al.[26] noted that in *Ezo epiyessoense* (Lithophylloideae) sexual plants were at least as common as asexual plants and that the latter produced only uninucleate bispores. They suggested that diploid tetrasporophytes in which meiosis occurs might be present in seasons other than when the collections were made.

Bisporangial plants are common in the Corallinaceae. For example, about 75% of the numerous specimens of *Bossiella orbigniana* examined in a monograph were bisporangial.[227] Only bisporangial plants of *Lithothrix aspergillum* occur north of Point Conception, Ca., but bisporangial, tetrasporangial, and sexual plants occur in southern California.[185,188] In this species, bisporangia and tetrasporangia sometimes occur together within the same conceptacle. Bauch,[37] noting the high frequency of bisporangia in the Corallinaceae, suggested that there are two types of species: (1) those that are obligate in producing bisporangia and (2) those that are facultative in this regard. In the latter type, some species would have distinct bisporangial and tetrasporangial strains, possibly related to locality, but in other species bisporangia and tetrasporangia would occur in the same plants, and sometimes even within the same conceptacle.

Suneson[427] provided cytological evidence on coralline life histories in work on *Dermatolithon litorale* and *D. corallinae* (both as species as *Lithophyllum*). The former was found to be an obligate bisporangial species, with meiosis absent and the resulting spores uninucleate (Figure 6, Chapter 1). Sexual plants were not present. Occasionally Suneson found abnormal reproductive cells such as binucleate tetrasporangia in which three cleavages occurred instead of the usual one.

Dermatolithon corallinae, an epiphyte of *Corallina officinalis* in much of Europe and elsewhere, was found to be a facultative bisporangia producer.[427] In the Mediterranean Sea only tetrasporangial plants were found. But in northern Europe, such as in Sweden, sexual individuals (usually monoecious) as well as plants containing binucleate bisporangia were found. In addition, Suneson also found in these northern strains, tetrasporangia and four-nucleate bisporangia intermingled in the same plants. Furthermore, his cytological evidence, based on feulgen-stained material, revealed that meiosis occurred in four-nucleate sporangia whether they became bisporangia or tetrasporangia. Nuclear divisions in binucleate bisporangia were apomeiotic. According to Dixon,[135] "This is the only clear-cut instance of facultative apomeiosis in the Rhodophyta for which the supporting data are sufficiently sound."

In northern Europe, tetraspores or two-nucleate bispores probably germinate into sexual plants whereas uninucleate bispores perpetuate asexual plants. In the Mediterranean, the life history of *D. corallinae* is probably of the usual *Polysiphonia* type, without the intervention of bisporangia.

SUMMARY

Reproduction in coralline algae is mostly by the formation of tetraspores (or bispores), spermatia, carpogonia, and carpospores within conceptacles.

Vegetative reproduction is probably common in coralline algae, but little is known about how this occurs. Potential dispersal units have been reported in *Jania* and *Fosliella*. Coralline fragments probably continue to grow even though broken away from attached plants, (e.g., marl). Attachment discs serve to anchor articulated thalli to hard surfaces.

Conceptacles are roofed chambers opening by one or several pores connected to chambers by canals. The types of conceptacle development depend on the taxa and the contained reproductive cells. They are borne on calcified crusts, protuberances or intergenicula. In most taxa primordia originate in meristematic tissue and result in an excessive cuticular secretion to form a conceptacle cap.

There are four types of development of tetrasporangial conceptacles in coralline algae. The types are categorized by whether the roofs form by circumconceptacular growth or by the growth of filaments among the tetrasporangia. In most taxa the conceptacles open by a single pore, but in the Melobesioideae a pore forms in the space occupied by a gelatinous plug formed at the apex of each sporangium. In *Sporolithon* a single tetrasporangium occupies each chamber.

Tetrasporangia are large and zonate in almost all taxa. Post-meiotic nuclei are extremely active in synthesizing cell material.

Male conceptacles exhibit a diversity of shapes, but most of them develop when surrounding filaments grow centripetally over the fertile area. Different developmental types occur in the Melobesioideae, some of them involving an origin below the intercalary meristem.

Male cells are produced in huge numbers on fertile surfaces consisting of basal cells, spermatangial mother cells, spermatangia, and possibly spermatia.

Carpogonial filaments are produced in uniform layers on the floors of female conceptacles. They are usually two-celled, with those in the center of a conceptacle producing long trichogynes that project through the pore. One to three carpogonial filaments are produced per supporting cell.

Following supposed fertilization zygotic nuclei pass into supporting cells via the fused cells of carpogonial filaments or by means of transfer tubes that bypass the hypogynous cells. Various kinds of fusion cells are formed, mostly from supporting cells. Gonimoblast filaments producing large unicellular carposporangia grow out from the margins or upper surfaces of the fusion cells. In some members of the Melobesioideae gonimoblast filaments fuse with other cells as they grow.

Coralline algae typically have a *Polysiphonia* type of life cycle, although this has not yet been completely demonstrated. Some species produce bispores which may be a way of surviving at the edges of the geographic range. Possibly uninucleate bispores form when meiosis fails to occur.

Chapter 6

CALCIFICATION

INTRODUCTION

Coralline algae are prominent among marine plants for their ability to deposit large amounts of calcium carbonate in their cell walls. The mineral is deposited as calcite with high concentrations of magnesium present. In fact, coralline algae are the only marine macroalgae depositing calcite; other calcifying red algae, as well as some green algae, deposit aragonite. Of great interest are the mechanisms by which biomineralization occurs. However, the paucity of information on the biochemical and physiological aspects of calcification in coralline algae becomes evident when recognizing that in a review on silicification and calcification Darley[107] devoted less than one page to these plants.

Several recent reviews deal with carbonates in the macroalgae, and these are drawn upon extensively. Milliman[320] considered all aspects of carbonate production from the point of view of a geologist. Important papers have also been contributed by Borowitzka[45] and Digby.[134,134a,134b]

The two crystal types or polymorphs of calcium carbonate in marine plants are calcite, a hexagonal rhombohedral form, and aragonite, an orthorhombic form. The needle-like aragonite is produced in intercellular spaces in certain green (Chlorophyta) and noncorallinaceous red algae. Calcifying green algae include *Halimeda*, *Penicillus*, *Rhipocephalus*, and *Udotea* in the *Caulerpales*, and *Acetabularia* and *Cymopolia* in Dasycladales. The only brown algal genus containing calcifying algae is *Padina* (Dictyotales).

Aragonite producers in the red algae include *Galaxaura* and *Liagora* (Nemaliales) and *Peyssonellia* (Cryptonemiales). Among marine plants the only calcite producers are the planktonic coccolithophorids (Prymnesiophyceae) in which the crystals are formed intracellurlarly within Golgi bodies, and the Corallinaceae where they form in the cell walls. The aragonitic algae do not become as solidly calcified as do the coralline algae. As shown in Figure 1, the cells of coralline algae are completely encased in fused calcite crystals except for the plugs of the primary pit-connections at each end. The crystals are organized with their c-axes at right angles to the cell lumen in the middle wall layer and are randomly oriented closer to the middle lamella.[33,49]

CELL WALL COMPOSITION

Milliman[320] has summarized reports of elements other than calcium in carbonates produced by marine organisms. Coralline algae contain the cations magnesium, iron, manganese, sodium, strontium, potassium, and barium. These cations may be incorporated into an organic matrix, be part of other mineral phases, or be "captured sea salts, such as NaCl or KCl."[320]

Generally, they are probably incorporated into the calcium carbonate lattice structure.[320] The tendency is for aragonite to concentrate cations with large ionic radii and calcite those with small ionic radii. Hence, magnesium, manganese, and iron tend to be more plentiful in calcite. Higher strontium values are usually reported for aragonite. Coralline algae, as well as many benthic foraminifera, sponges, octocorals, bryozoans, decapods, and echinoderms, usually contain more than 8 mol % magnesium carbonate, that is, 2% magnesium by weight. Coralline algae are major producers of magnesium calcite in the oceans, especially in tropical areas.

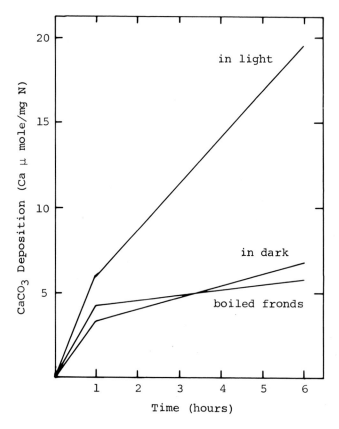

FIGURE 1. Incorporation of Ca[45] into calcium carbonate in *Serraticardia maxima* from Japan. Adapted from Okazaki, et al.[342]

The quantification of carbonate mineral constituents has been considerably advanced by the use of X-ray diffraction.[320] Calcites produced by different organisms vary considerably in peak heights.[320] Probably the best use of this technique has been to determine magnesium content. Milliman gave mole percent magnesian calcite for several coralline algae from shifts in the X-ray diffraction peaks. In the algae the mole percents ranged from 9.5 in *Lithothamnium* to 20.0 in *Lithophyllum*, and even more than 25 in other tropical entities. These amounts contrast with most animal calcites in which the magnesium content is less. In fact, Milliman et al.[323] found excess magnesium (more than 4-6% Mg) in the tropical *Porolithon*, *Neogoniolithon*, and *Amphiroa* (Figure 2) and suggested that part of it is present as magnesian calcite and part as brucite [$Mg(OH)_2$].

Brucite was first verified as present in coralline cell walls in *Goniolithon* (= *Neogoniolithon*) by Weber and Kaufman[443] and Schmalz.[369] Unexplainable discrepancies in the carbonate fractions suggested to these workers and others that more than magnesian calcite is present. It is difficult to explain why brucite is present. The question that must be asked is: Is it a characteristic of the species examined, or a reflection of ambient environmental conditions, or are there unknown reasons for its presence in coralline cell walls?

THE CALCIFICATION PROCESS

Prerequisites for the deposition of calcium carbonate include having appropriately high concentrations of calcium and carbonate as well as an alkaline environment.

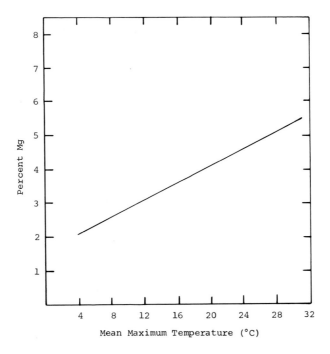

FIGURE 2. Magnesium concentrations as a function of temperature in various specimens of coralline algae. Adapted from Milliman, et al.[123]

Tropical waters are supersaturated with calcium and carbonate, the latter mostly as bicarbonate ions, whereas in cold waters these occur in somewhat lower concentrations.[320] If the concentrations of these ions can be slightly increased in the cell walls where accretion is to occur and if the pH can be made sufficiently high, deposition is likely to take place. Borowitzka[45] summed up the data available on calcification processes while also showing that for coralline algae there are many unanswered questions.

One area of general agreement is that light stimulates calcification rates (Figure 2),[47,342,467] and that photosynthesis is closely related to the process.[134,134a] Pentecost[473] found that in *Corallina officinalis* from Great Britain calcification rates in apical intergenicula in the dark were appreciable; 30 to 40% as great as in the light (at saturation). The light to dark calcification ratio was about 3 in apical intergenicula and 3.5 to 4 in lower intergenicula. In apical intergenicula of *C. officinalis*, the mean ratio of photosynthetic[14] CO_2 uptake to incorporation of $^{14}CO_2$ into calcite was 1.03.[467] In the dark, calcification is probably due to at least two effects: (1) a delayed light-dependent reaction and (2) a physical absorption process. Borowitzka[47] demonstrated that in *Amphiroa foliacea* Ca^{45} exchanges at three sites (in three compartments). Two sites appear to represent organic binding and the third isotopic exchange on calcium carbonate crystal surfaces. Pearse,[349] in studies of *Bossiella orbigniana*, pointed out that although stimulation of calcification is apparently related to photosynthesis, the light effect is greatest at branch apices where there is the least photosynthetic pigment. She suggested that photosynthetic products translocated from lower intergenicula may serve as components of organic matrices or as energy sources in the apical intergenicula. Other reports indicate that in young intergenicula rates of both calcification and photosynthesis are higher than in older parts of the thalli.[47]

Goreau,[190a,191] working in Jamaica, determined calcification rates from calcium-45 incorporation after it was added to the seawater medium as calcium-45 carbonate under various environmental conditions. He found that for the 15 species he studied red algae are faster calcifiers than green algae. More than one half of the plants had light to dark ratios greater than one. Most of the algae with ratios greater than two were restricted to waters deeper than 25 m. However, *Amphiroa fragilissima* was unusual in that it calcified more slowly in the light than in the dark.[190a] Goreau[190a] hypothesized that light inhibition of calcification in this species is because of a shortage of available bicarbonate due to competition for carbon dioxide as a common substrate for the high level of photosynthesis. He suggested two ways in which light may stimulate calcification:

1. A direct controlling of bicarbonate concentrations in tissues by photosynthesis.
2. An indirect augmentation of the supply of free energy available for active calcium and bicarbonate transport.

Recent support of this was the finding by Okazaki[340] that calcium-dependent ATPase activity was relatively higher in three species of articulated coralline algae than in several noncorallinaceous marine algae. Based on this he speculated that ATPase near or in the cell membrane may free energy from ATP produced during photosynthesis, and that the energy might be used in active ion transport to the site of calcification.

In a review of algal calcification, Borowitzka[45] discussed three theories which have been proposed to explain carbonate deposition. In the first theory, the "Carbon Dioxide Utilization Theory," the assumption is that carbonate deposition is a consequence of photosynthetic extraction of carbon dioxide from water during photosynthesis. In the equation below this would result in the equilibrium shifting to the right:

$$Ca^{2+} + 2HCO_3^- \leftrightarrow CaCO_3 + H_2O + CO_2$$

As pointed out in several papers, the most serious drawback to this idea is that noncalcareous algae growing next to calcareous algae do not calcify. Goreau,[191] in studies mostly of hermatypic coral animals, suggested that carbonic anhydrase may be important in calcification. However, Okazaki[339] has found that this enzyme is no more active or more concentrated in *Serraticardia maxima* than in noncalcareous algae. Goreau[475] summarized this theory in the four equations below, of which carbonic anhydrase catalyzes the last one:

$$Ca^{2+} + 2HCO_3^- \leftrightarrow Ca(HCO_3)_2$$

$$Ca(HCO_3)_2 \leftrightarrow CaCO_3 + H_2CO_3$$

$$H_2CO_3 \leftrightarrow H^+ + HCO_3^-$$

$$H_2CO_3 \leftrightarrow CO_2 + H_2O$$

A newer idea is that a protein — polysaccharide complex acts as a site for carbonate deposition; this is Borowitzka's[45] "Organic Matrix Theory." According to this hypothesis, control by the organic substance would be manifested by concentrating calcium or bicarbonate ions, or by forming a template for carbonate nucleation. This mechanism would explain why calcite instead of aragonite is precipitated in coralline algae. Borowitzka[45] suggested the possibility that cellulose, or a complex including cellulose, serves this function. In fact, the intimate relationship between organic and inorganic components is evident in published electron micrographs showing crystals within an organic matrix.[50,50a]

In support of the matrix hypothesis, it is interesting to contrast accretion in aragonitic green algae and coralline algae. Borowitzka et al.[49] pointed out that the precipitation of aragonite in green algae resembles simple precipitation more than it does in the coralline algae. In the calcitic coralline algae, the crystals are placed in an intimate association with an organic cell wall. Since the nature of various organic substances has been shown to determine whether the deposition is calcitic or aragonitic, it is obvious that this matrix in coralline algae favors the deposition of the former. It could be that the function of a matrix in calcite crystallization is interfered with by phosphate. The lowering of phosphate concentrations below the usual levels in laboratory culture has been found necessary for optimal growth of some coralline algae,[58,420] There is a strong link between calcification and other metabolic processes and, according to Borowitzka et al.,[49] the coralline cells maintain close control over the formation of the crystals. This is in contrast to aragonitic green algae where a mechanism initiates crystallization which proceeds in a disorganized fashion with the influence of the alga being to produce a slightly more favorable environment for precipitation.

The "Bicarbonate Usage Theory"[45] states that bicarbonate is taken up for photosynthesis and that hydroxyl ions produced maintain a basic pH conducive to calcification. Strong support for this comes in two recent papers where Digby[134,134a] described some experiments on calcification that he carried out on *Clathromorphum circumscriptum* and *Corallina officinalis* from Maine. In his second paper he presented an hypothesis outlining possible mechanisms in the accretion of calcium carbonate by coralline algae. The results that Digby obtained may be summarized as follows:

1. Growth rates in *C. circumscriptum* are slow, with a yearly increase in thickness, in exposed and well illuminated sites, of about 0.3 mm.
2. Diffusion from exposed inner regions of a thallus is relatively greater than from an intact surface. Damage results in a local softening.
3. The surfaces of living thalli are slightly acidic in relation to areas just below, where the conditions are more alkaline.
4. Supernatants from ground-up thalli were more alkaline if the plants were exposed to light than if they had been kept in the dark.
5. The vacuolar sap is acidic.
6. Living plants in clear tubes set out in the field turned the water alkaline. In the dark the water became slightly acidic.
7. Oxygen produced in *C. circumscriptum* in bright sunshine was well in excess of the equivalent weight of carbonate accretion. In the dark the rates were about equal.

The hypothesis that Digby[134a] presented places emphasis on the chemical events as controlled by photosynthesis and pH. A great proportion of the energy from photosynthesis is presumably available for carbonate secretion considering the fact that growth is very slow and organic production is low, especially in cold parts of the world such as the coast of Maine. The scheme outlined below and in Figure 3 is based on the possible reactions occurring in and around a single calcifying meristematic cell.

After storage in unknown intermediate compounds, the energy produced in photosynthesis is channeled into the secreting cell where it serves to prepare carbonate for precipitation with calcium from the ambient seawater. Interaction with photosynthesis is also such that the calcifying region of the thallus and the intracellular environment is made appropriately alkaline, a condition prerequisite to carbonate deposition. This establishment of an alkaline environment is particularly important in cold waters such as in Maine, where the seawater is slightly undersaturated with calcium carbonate. In the scheme that Digby proposed, he pointed out that the alkaline environment must be balanced by an equal production of acid. His hypothesis accounts for this by pos-

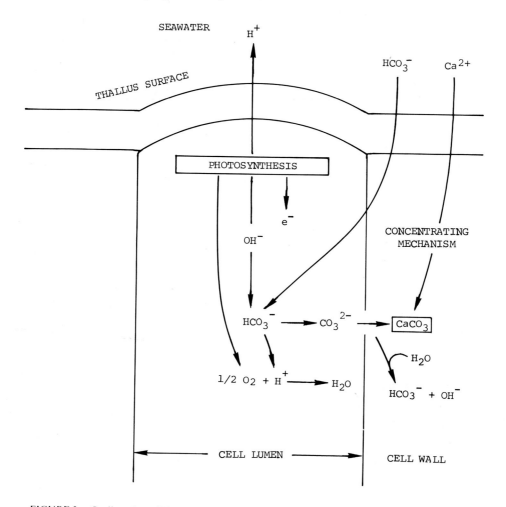

FIGURE 3. Outline of possible mechanisms for calcification in coralline algae. See text for details.

tulatng a secretion of hydrogen ions from the noncalcifying apices of the terminal thallus cells. Digby's scheme is based on the assumption that bicarbonate ions from the sea are precipitated by the addition of one equivalent of base:

$$Ca^{2+} + OH^- + HCO_3^- \rightarrow CaCO_3 + H_2O$$

In the following steps in Digby's hypothesis the first is the photosynthetic splitting of water:

$$H_2O \rightarrow \tfrac{1}{2} O_2 + 2H^+ + 2e^- \qquad (1)$$

Oxygen atoms and hydrogen ions pass from the cell and the electrons are retained, presumably reducing organic metabolites. As these metabolites are oxidized hydroxyl ions are produced in excess due to reduction of the hydrogen ions by electrons during photosynthesis:

$$H_2O \leftrightarrow OH^- + H^+ \rightarrow \qquad (2)$$

$$H^+ + e^- \rightarrow H\cdot$$

Other electrons originally derived from photosynthesis react with bicarbonate ions that have diffused in from the sea to form carbonate ions:

$$2HCO_3^- + 2e^- \rightarrow 2H\cdot + 2CO_3^{2-} \qquad (3)$$

The presence of hydroxyl ions would also result in the conversion of bicarbonate to carbonate in a reaction possibly involving carbonic anhydrase:

$$OH^- + HCO_3^- \rightarrow H_2O + CO_3^{2-} \qquad (4)$$

The next step postulates the outward diffusion of carbonate ions and their internal replacement by inwardly diffusing bicarbonate. Outside the cell the carbonate ions partly hydrolyze, thus raising the pH in cell walls where carbonate is being deposited:

$$2CO_3^{2-} + H_2O \rightarrow 2HCO_3^- + 2OH^- \qquad (5)$$

The conditions of saturation of calcium and carbonate ions are reached in the walls with the result that calcium from the seawater and the carbonate from the cells form calcium carbonate:

$$2Ca^{2+} + 2CO_3^{2-} \rightarrow 2CaCO_3 \qquad (6)$$

Finally, the hydrogen atoms which appeared earlier (reactions 2 and 3) must be removed by combination with oxygen:

$$2H\cdot + \tfrac{1}{2} O_2 \rightarrow H_2O \qquad (7)$$

Digby pointed out that with the diffusion of hydrogen ions outward through the tips of the cells, and the inward diffusion of bicarbonate ions and outward diffusion of carbonate ions at the sides of the cells, calcification would occur. The hydrogen ions passing out through the cell apex would be lost to the surrounding sea. This scheme could account for carbonate precipitation on the sides of growing cells only micrometers away from the tips of the cells where conditions would not be alkaline enough for precipitation. Probably this scheme is an oversimplification and more steps or refinements will be discovered. It does form a plausible framework which does not involve simple removal of carbon dioxide by photosynthesis and does not exclude the idea than an organic matrix in the cell wall plays an important role in deposition.

SUMMARY

Surprisingly little is known about the process of calcification in coralline algae. These plants produce calcite, in contrast to other benthic algal calcifiers that precipitate aragonite.

Several cations have been detected in the calcium carbonate of coralline algae, but the most prominent is magnesium. Magnesium is especially precipitated in warm waters, and probably occurs as magnesian carbonate and, at least in one species, as brucite.

Increasing the concentrations of calcium and carbonate in the cell walls and establishing an alkaline environment fosters biomineralization in coralline algae. Deposition is usually faster in the light than in the dark, and photosynthesis is implicated. However, calcification is rapid in unpigmented thallus margins and apices and probably energy-rich compounds are translocated to the sites of biomineralization. Three theories have been presented for the process of calcification: (1) carbon dioxide utilization, (2) organic matrix, and (3) bicarbonate usage. The mechanism of calcification in coralline algae probably involves elements of the second and third theory.

Chapter 7

PHYTOGEOGRAPHY

INTRODUCTION

The phytogeographical treatment in this chapter will, of necessity, emphasize distribution in certain areas and because information is not available or is unreliable by modern standards, other areas will be notably absent or treated only briefly. Distributional studies fall into two categories: (1) descriptive and (2) causal. Very little can yet be said about causal studies, but Adey's studies in the North Atlantic are pioneering in this regard. For other vast areas, all that can be done is to give rough descriptions of which coralline algae are present, and even here there are huge gaps. Ideally, the oceans and bordering lands could be divided into 12 major regions (Table 1, Figure 1), and for each the coralline distributions could be described. This would be an enormous task, but one that will in all likelihood be done little by little as future studies are made. These suggested regional units are delimited approximately by political boundaries when possible. The divisions resemble those given by Littler.[284]

Adey[10] presented a series of world maps on which he summarized the relative distributional importance of several crustose coralline genera. These maps, which were republished by Adey and MacIntyre,[24] give an approximate idea of the worldwide distribution of these genera based on Adey's interpretation of Foslie's many species (Table 2). This first attempt at understanding coralline distribution on a world scale has some deficiencies. For the most part, only Foslie's species were considered and up-to-date identification data are lacking for much of the world. Still, this is a useful approach and reveals that some genera are predominant in the cold northern hemisphere (e.g., *Lithothamnium* and *Clathromorphum*), while others predominate in the cold southern hemisphere (e.g., *Mesophyllum*), and still others are tropical (e.g., *Neogoniolithon* and *Porolithon*).

In certain parts of the world, particularly in upper subtidal zones in temperate and subtropical areas, nearly pure stands of articulated coralline algae may lend a pink cast to the shore. Examples of this may be found in New England and northern Japan (*Corallina*), in kelp forests in California (*Calliarthron*), in eastern South Africa (*Arthrocardia*), subtidally in southern Australia (*Metagoniolithon*), and possibly in other areas of dense population. Although not much is known about why these plants grow where they do, ecological information is accumulating for some taxa in certain areas and will be considered later. General aspects of the distribution of the genera of articulated coralline algae are given in Table 3.

COLD NORTHWESTERN ATLANTIC

Adey has made extensive studies of the distribution, ecology, and taxonomy of epilithic crustose coralline algae in the Gulf of Maine.[3,5,6-8,11] Most of what follows comes from his papers. Some of the advantages of following Adey's papers are that he has (1) studied and rigorously followed the taxonomy and type specimens set forth by Foslie at the beginning of this century, (2) examined large numbers of specimens utilizing a population approach, and (3) studied crustose coralline algae at numerous stations in the northeast.

The epilithic crustose coralline algae present on the east coast of North America are dominated by the Melobesioideae. Identification may be made on the basis of repro-

Table 1
MAJOR PHYTOGEOGRAPHICAL REGIONS

REGIONS	LIMITS
1. Cold NW Atlantic	South to Cape Hatteras (35°N. Lat.)
2. Cold NE Atlantic	South to Straits of Gibralter (about 35° N. Lat.)
3. Cold NW Pacific	Above 30° N. Lat.; includes Japan
4. Cold NE Pacific	South to Point Conception, Calif. (35° N. Lat.)
5. Warm W Atlantic	South to Cape Hatteras (35° N. Lat.) and north of the Uruguay-Argentina border (35° S. Lat.)
6. Warm E Atlantic	The west coast of Africa from the Straits of Gibraltar (35° N. Lat.) to the Cape of Good Hope (about 35° S. Lat.)
7. Warm W Pacific	Between 30° N. Lat. and 30° S. Lat. and eastward to include French Polynesia (120° W. Long.)
8. Warm E. Pacific	Between Point Conception (35° N. Lat.) and the Peru-Chile border (20° S. Lat.)
9. Indian Ocean	Including eastern Africa from the Cape of Good Hope to the Bay of Bengal
10. Mediterranean and Red Sea	
11. Southern Australia and New Zealand	Below 30° S. Lat. in Australia
12. Antarctic and Subantarctic	Below 50° S. Lat., but also including Chile and Argentina

ductive, vegetative (meaning sectioned specimens not needed), and anatomical characteristics. Keys to the species are given in Adey[11] and Adey and Adey.[20]

Particularly important in the western part of the North Atlantic are two species of *Clathromorphum*, *C. circumscriptum* and *C. compactum* (Table 2, Chapter 4). No species of *Clathromorphum* occur in Great Britain.[20] *C. circumscriptum* occurs in the subarctic North Atlantic as well as in the North Pacific.[262] In the western Atlantic, it has been reported as far south as Cape Cod, and it is the dominant crustose species in

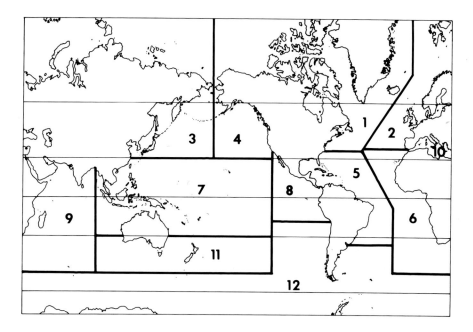

FIGURE 1. World map showing the 12 geographic regions described in the text.

the upper 1 m in much of Newfoundland and Labrador. In the more southern parts of its range, it tends to grow in the upper subtidal zone, but in colder areas it extends not only upward into intertidal pools but also into deeper water.[5] The second species in the North Atlantic, *C. compactum*, is less widespread and has never been found in the intertidal zone.[5] The southernmost record is from southwestern Maine, and from here it increases in frequency into colder waters, gaining its greatest abundance deeper than *C. circumscriptum*, at a depth of 6 to 9 m in the mouth of the Bay of Fundy. It is absent where summer temperatures are higher than 15°C.

Phymatolithon laevigatum requires warmer water than the other species in this genus, being the main crustose species just south of Cape Cod and distributed in warmer waters north to the Gulf of St. Lawrence. The well-studied species *P. lenormandii* (usually listed as *Lithothamnium lenormandii*) is essentially an intertidal species common in the Gulf of Maine from Casco Bay to Nova Scotia, where it often grows under the fucoid *Ascophyllum nodosum* (Linnaeus) Le Jolis.[8] It is often found with *P. laevigatum*. Its range is boreal to subtropical although its distribution is somewhat erratic.[20] Unlike the other northern species of *Phymatolithon* that produce conceptacles in winter, *P. lenormandii* produces them year-round. In the Gulf of Maine Adey[8] has always found gametophytes to be hemaphroditic, with spermatangial filaments within female conceptacles. The most widely distributed boreal crustose coralline species is *P. rugulosum*, which is a relatively deep-water plant[20] and can be found anywhere from the Gulf of Maine to the Gulf of St. Lawrence.[6]

The genus *Lithothamnium* is abundant in the North Atlantic. *Lithothamnium glaciale* (referred to as *Lithothamnium* "a",[8,11]) is possibly the most abundant crustose species in subarctic waters.[20] From Cape Cod at its southern limit, it becomes more and more prominent farther north.

Probably the only species of *Lithothamnium* that is endemic to Atlantic North America is *L. lemoineae* (referred to as *Lithothamnium* "b."[8,11]) Like some of the other crustose coralline algae in this area, it reproduces in winter and, like *Clathromorphum circumscriptum*, it may require temperatures below 2°C to produce concep-

Table 2
AREAS OF GREATEST CONCENTRATION OF NONARTICULATED SPECIES OF CORALLINE ALGAE[13]

GENERA	AREAS OF GREATEST CONCENTRATION
Clathromorphum	Cold northern hemisphere
Goniolithon	Tropics
Hydrolithon	Tropics
Leptophytum	Cold northern and southern hemispheres, absent in tropics
Lithophyllum	Tropics
Lithoporella	Tropics
Lithothamnium	Cold northern hemisphere, deep in tropics
Mesophyllum	Cold southern hemisphere
Metamastophora	Tropics
Neogoniolithon	Tropics
Phymatolithon	Cold northern hemisphere
Porolithon	Tropics
Pseudolithophyllum[a]	Cold southern hemisphere
Sporolithon	Tropics

[a] As interpreted by Adey.[13]

tacles.[12] This species is prevalent in the Gulf of Maine and extends up to Baie des Chaleurs,[80] but in diminishing amounts. Only bisporangial plants have been reported.[8] *Lithothamnium tophiforme* (referred to as *Liththamnium* "c",[8,11]) is also a deep-dwelling entity usually found below 15 to 20 m but, in contrast to *L. sonderi*, this species is arctic in distribution. It has been collected from Newfoundland and northward.

Only two species of *Leptophytum* have been described, and both of them occur in the North Atlantic. They are thin (<500 μm) crustose species, without conspicuous surface irregularities and with plants rarely overgrowing one another.[8] The two species are difficult to distinguish from some species of *Phymatolithon* and, in fact, the two genera are closely related. *Leptophytum foecundum* is a crustose species with its center of distribution in the arctic. It occurs southward to the Isle of Shoals (off New Hampshire), but is restricted to deep water in the Gulf of Maine. It has been recorded from northern Newfoundland, Labrador, and is common in eastern Greenland.[292] Rarely discovered off northern Iceland, it has not been reported from Great Britain. The other

Table 3
GENERAL DISTRIBUTION OF THE GENERA OF ARTICULATED CORALLINE ALGAE[223]

GENERA	AREAS OF DISTRIBUTION
Alatocladia	Japan
Amphiroa	Worldwide in tropics and subtropics
Arthrocardia	**Southern Africa,** India, Brazil, Northern California
Bossiella sg. *Bossiella*	**Eastern Pacific** from Alaska to South America and to Argentina in South Atlantic
Bossiella g. *Pachyarthron*	Western and northern Pacific from Japan to Alaska and in the eastern Pacific south to Canada; possibly in the southern hemisphere in Chile and S. Australia
Calliarthron	Western Pacific in Japan; eastern Pacific from Alaska to Baja California
Cheilosporum	**Subtropical:** Indian Ocean, Brazil, S. Japan
Chiharaea	**Eastern North Pacific** in British Columbia and Northern California
Corallina	**Cosmopolitan** except for tropics
Haliptilon	Tropical and **subtropical** and warmer parts of Europe
Jania	Worldwide in **tropics** and **subtropics**
Lithothrix	**Eastern North Pacific** from British Columbia to Baja California
Marginisporum	Japan
Metagoniolithon	Southwestern and southern **Australia**
Serraticardia	Western North Pacific in Japan; eastern North Pacific from Alaska to southern California
Yamadaea	Western North Pacific in Japan; eastern North Pacific in Washington and Central California

Note: The main areas of distribution are in boldface type where pertinent.

species of this genus, *L. laeve,* is abundant in arctic and subarctic areas in deep water, especially on shell and pebble bottoms.[20] It occurs off Labrador and, together with *Lithothamnium glaciale,* is the dominant coralline species in deep waters of the Gulf of Maine.[11]

In the western North Atlantic *Lithophyllum orbiculatum* is present from Cape Ann, Mass. to Newfoundland, being very abundant in the Bay of Fundy.[7]

From Cape Hatteras, N. C. and into the arctic only one articulated species, *Corallina officinalis,* has been reported. This, the first articulated coralline species to be given a binomial,[280] has been cited in floristic works from many parts of the world, but it remains to be seen how widespread it really is. It is absent from tropical regions, as is the entire genus *Corallina* (not to be confused with those species of *Corallina* that need to be transferred to *Haliptilon*).

COLD NORTHEASTERN ATLANTIC

In this region, extending north of the Straits of Gibraltar (about 35 N. Lat.), the nonarticulated coralline algae are better known than anywhere else in the world. Adey's studies in Iceland,[9,10] Norway,[14] and Great Britain[20] have provided much distributional information on epilithic genera (Table 4). The species occurring as marl have been studied by Adey and McKibbin[27] and Cabioch.[64,68,73] Chamberlain[84-86,88] has published information on epiphytic crustose coralline algae and there are numerous other papers containing geographical information for this part of the world.

The cold water genus *Clathromorphum* is represented by two species in the North Atlantic, *C. compactum* and *C. circumscriptum.* The former is known from northern Europe, the U.S.S.R., and the northern Pacific.[5,262] *C. circumscriptum* occurs south to Trondheimsfjord, Norway (about 71 N. Lat.) and is common in northern and eastern Iceland.[14]

Phymatolithon is well represented in the eastern North Atlantic with five known species. *P. laevigatum* occurs on the south coast of Iceland and is the most abundant shallow water crustose coralline at Trondheim, Norway.[14] In the British Isles it occurs in bays, but it has not been reported south of the English Channel. From its distribution, it is possible that this species is tolerant of low salinities, such as is *P. polymorphum. P. polymorphum,* a species absent in the western Atlantic, is the dominant crustose coralline species in wave-exposed areas of central Europe in shallow to somewhat lower levels.[14] It is very abundant in Great Britain and occurs at least as far south as Morocco[197] and up to northern Norway and probably into the U.S.S.R. It is absent in the western North Atlantic, but important in southern Iceland. Although usually an epilithic crust, it forms marl in places like Galway Bay and nodules 20 cm in diameter have been found in northern Norway.[20] It has been shown to produce sexual conceptacles in winter and, like *P. laevigatum,* probably requires winter temperatures that do not go below 1 to 2°C. Another species, *P. lenormandii,* is primarily intertidal in fjords of southern Norway, in southern Iceland, and in the British Isles.[20] The deep water *P. rugulosum* has been collected off Iceland and in mid-depths off Great Britain. These species of *Phymatolithon* serve to make this genus the dominant crustose coralline genus in the boreal North Atlantic. Of the five species above, *P. calcareum* is aberrant in that it is the only northern species of *Phymatolithon* that is primarily a marl-former. Also, its range of distribution is such that it extends into the northern North Atlantic only as far as Great Britain, southern Norway (although Adey[14] stated that he did not find this species in Norway), and mainland Europe.[20]

Among the crustose coralline algae the genus *Lithothamnium* dominates in much of the arctic, occurs only in the deep water in the tropics, and apparently does not inhabit Antarctic waters (Table 2). In the Atlantic above 40° N. Lat., five species are present.[20]

Table 4
NUMBERS OF SPECIES OF CORALLINE ALGAE REPORTED FROM THREE AREAS OF THE NORTH ATLANTIC[a]

	Gulf of Maine	Iceland	Great Britain
Choreonema	—	—	1
Clathromorphum	2	1 (2?)	—
Corallina	1	1	2
Dermatolithon and Tenarea	1	1 (?)	8
Fosliella	2	1 (?)	7
Haliptilon	—	—	1
Jania	—	—	2
Kvaleya	1	—	—
Leptophytum	2	2	1
Lithophyllum	1	1	4
Lithothamnium	3	2 (3?)	1
Melobesia	1	(?)	2
Mesophyllum	—	—	1
Phymatolithon	3	4	6
Schmitziella	—	—	1

[a] Derived mostly from papers by Adey and his co-workers and from Parke and Dixon.[348]
[b] — indicates not recorded.

Four of them are truly epilithic crusts, although some of them may form marl, whereas *L. coralloides* is usually found as marl. *Lithothamnium sonderi* is a deep-growing plant that Adey[14] suggested cannot survive winter temperatures below 6 to 7°C. It is the main crustose species in deep waters (often below 15 m) off Great Britain and southern Norway; it is absent in North America and Iceland and perhaps occurs sparsely in the Mediterranean.[197] The well-known species *Lithothamnium glaciale*, which is also present in the western North Atlantic, occurs in northern Great Britain and becomes more and more dominant in northern Norway and in Iceland except for the southern coast. Commonly attached in deep water, it also may be the main marl in the subarctic, being partly replaced by *L. tophiforme* in the arctic.[20] Adey[14] noted that it grows well in waters up to 18 to 19°C and that sexual reproduction occurs in the winter and spring. Thus, he suggested that the critical low temperature for conceptacle formation is 4 to

5°C. *L. tophiforme* occurs in northern Iceland, northern Norway, Greenland,[292] and probably in the U.S.S.R. It does not occur in Great Britain, possibly being limited in its southward distribution by maximum temperatures of 8 to 10°C.[12]

Leptophytum, represented in this part of the world only by the deep-water *L. laeve,* occurs off Great Britain, Norway, and Iceland.[20]

Three species of *Lithophyllum* occur in the northern North Atlantic, but only one of them, *L. orbiculatum,* extends into the subarctic. *L. incrustans* is the major encrusting coralline in tide pools and shallow water in northwest Spain, and it extends southward at least to Cape Blanco.[20] It is also common in the British Isles, especially in tidepools, but disappears in Norway. It has not been reported from Iceland or the western North Atlantic. Only recently described,[20] *L. nitorum* is not a cold water species, occurring in the Bay of Biscay, abundantly in Cote Basque and in the British Isles, usually in the mid-photic zone and below. It is only sporadic in northern Great Britain[20] and has not been reported from Norway, Iceland, or the western Atlantic Ocean. Adey[7] treated *L. orbiculatum* under *Pseudolithophyllum* until 1970[11] and relatively extensive distributional information is available for it. This species is widespread throughout the boreal and into the subarctic. It is especially important in northern and eastern Iceland,[9] Norway, and occurs in France, Spain, and Great Britain.[20]

Mesophyllum occurs primarily in the southern hemisphere (Table 2), and only *M. lichenoides* extends north of 35°N. latitude in the Atlantic Ocean. Like other species of *Mesophyllum,* this species is thin (100—620µm) and leafy, with uplifted margins, has a coaxial hypothallus, and large protruding conceptacles in which the chamber diameter is about 500 µm.[20] It is not epilithic in Great Britain but is sometimes so in France and Spain. It usually grows on *Corallina officinalis* and coralline debris.

Fosliella is probably widespread in the North Atlantic, and in Great Britain. It has recently been studied in detail.[20,84-85] Seven species were listed for Great Britain,[348] and the genus is relatively well known here, but elsewhere, especially in the western North Atlantic, very little information is available.[431]

Four genera of articulated Corallinaceae are present in the cold northeastern Atlantic. As may be seen in Table 4, *Corallina* is the most widely distributed, and may be considered a cold water entity. In contrast, *Jania, Haliptilon,* and *Amphiroa* are essentially tropical genera that do not extend far into the north.

Two species of *Corallina* predominate in Atlantic Europe: the ubiquitous *C. officinalis,* and *C. elongata* (cited in much literature as *C. mediterranea*). In addition, other species, such as *C. hemisphaerica* and *C. calvadosi,* have been described, but they have not been studied much and some of them may be conspecific with *C. officinalis* or *C. elongata. Corallina elongata* is more likely to be found in warmer waters than is *C. officinalis.* It is sometimes difficult to distinguish between these two species, but the shorter axial intergenicula, more compact branching, softer and more flaccid aspect, and the tendency for fused intergenicula to occur in branch endings should help to identify *C. elongata.*[197]

Two species of *Jania* are present in Great Britain[348] and, because one other species occurs in Portugal,[31] this genus is represented by three species in the Cold Northeastern Atlantic (Table 4). When recognizing that a profusion of poorly known species occur in warm waters of the world, it can be appreciated that the two species in Great Britain are those most adapted for cold water. The three species from the cold northeastern Atlantic are *J. rubens, J. corniculata,* and *J. longifurca.*

Haliptilon squamatum is a northward extension of a warm-water genus that reaches to western Ireland.[469,470] Distributional and reproductive studies of this species may reveal that its center of distribution is in Atlantic Spain and Portugal, or in the Mediterranean Sea.[197] Sexual plants occur in these areas as well as on the Atlantic coast of France, but perhaps only tetrasporangial plants are present in the British Isles. Webster[469] found *H. squamatum* in profusion on the lower parts of *Cystoseira* (Phaeophyta) in deep, well-aerated pools where occur the species of the "southern element"[106] in Galway and Ireland in summer. The coralline species was absent from this habitat in winter.

One species of the tropical genus, *Amphiroa*, *A. beauvoisii*, occurs in the cold North Atlantic, and that extends north only into Portugal.[31] Although monographic work on *Amphiroa* is needed, it appears that this species is the most widely distributed, having been reported from the western Indian Ocean,[224] the western North Atlantic below Cape Hatteras,[374] and possibly from both sides of the Pacific Ocean (Japan, as *A. zonata*),[455] and (Mexico, also as *A. zonata*).[113]

COLD NORTHWESTERN PACIFIC

Many species of nonarticulated coralline algae have been reported from Japan and nearby areas (e.g., Sakhalin and Korea) but, until monographic studies can be made, uncertainty exists as to the true identity of some of them. Masaki,[303] in a comprehensive morphological study of Japanese crustose coralline algae, described numerous species. More recent papers on coralline algae in Japan[26,262] suggest that monographic work would result in nomenclatural changes. Probably interesting floristic relationships between Japan and California will be revealed when studies can be made (Table 5).

Regarding seawater temperature and other climatic features, Japan is tropical in its southern parts and temperate to subarctic in the north. As a result, the diversity of coralline algae is large, with Masaki[303] including more than 40 nonarticulated species and Okamura[338] 38 articulated species. As in other areas in which few monographic studies have been made, those numbers are probably inflated, with some so-called species being growth forms of others. Furthermore, subtidal areas have not been examined for coralline algae in much of Japan. It must also be recognized that some of the names may have been incorrectly applied when type specimens or adequate descriptions were not available to authors of new species in the Japanese flora.

The relationships between seawater temperature and the distribution of crustose coralline algae in the waters around Hokkaido, the northernmost island of Japan, was recently studied by Adey et al.[26] They found that the large, current-influenced temperature variation around the island allowed for the establishment of three main groups of species, designated as the subarctic, cold temperate, and warm temperate groups (Table 6). The subarctic group included five species, four of which also occur in the subarctic North Atlantic. The fifth species, *Lithothamnium* "pac-lem," is closely related to the North Atlantic *L. lemoineae*. These species in northeastern Hokkaido represent a southward extension of a widely distributed subarctic group of crustose coralline algae. In Japan these species occur off eastern Hokkaido from Cape Erimo to Cape Nosappu where the depth distributions of the species generally match those distributions in the North Atlantic.[8] The waters off this part of the island are cold (surface temperatures of 0 to -1°C in winter and 12 to 15°C in summer) because of the southward flowing Oyashio current.

The cold temperate crustose coralline algae off Hokkaido consist of *Hydrolithon* sp. and *Lithothamnium* sp. These species are prevalent in northern Hokkaido from Cape Hosappu to Cape Shiretoko, where the seawater temperature is influenced by a summer warm current flowing north and east through Soya Strait. *Lithothamnium* is a widely distributed genus, although elements in tropical areas tend to grow in deeper water than in cold areas. *Hydrolithon* is tropical and as such appears out of place off northern Hokkaido.

The warm temperate crustose coralline flora extends from Tsugaru Strait to Point

Table 5
APPROXIMATE NUMBER OF SPECIES OF NONARTICULATED CORALLINE ALGAE REPORTED FROM JAPAN AND CALIFORNIA[a-b]

	Japan	California
Choreonema	1	1
Clathromorphum	3	2
Dermatolithon and Tenarea	4	2
Fosliella	3	1
Heteroderma	2	1
Hydrolithon	1	1
Leptophytum	1	—
Lithophyllum	10	5
Lithoporella	1	—
Lithothamnium	8	10
Mastophora	1	—
Melobesia	1	2
Mesophyllum	1	2
Neogoniolithon	3	1
Phymatolithon	1	—
Porolithon	3	—

[a] Derived from Masaki[303] and Abbott and Hollenberg.[1]
[b] — indicates not recorded.

Erimo in Southern Hokkaido. *Lithophyllum yessoense* and *Neogoniolithon* sp. dominate in this area, although *Mesophyllum* sp. is also present (Table 6). The boundaries of these floristic areas are not sharp, and particularly evident is the way that warm water species overlap cold water species, with the former in shallow water above the latter in deep water.

Japan has a diverse articulated coralline flora with most genera represented (Table 7). Absent are *Arthrocardia, Chiharaea, Lithothrix, Metagoniolithon,* and *Haliptilon,* although it is possible that a species in the last genus is present in southern Japan. One reason for the diverse array of articulated taxa is probably the fact that Japan extends from cold temperate regions in the north to subtropical regions in the south. Although two genera are endemic to Japan, it is noteworthy that Japan shares several genera and species with the eastern Pacific, particularly with entities from California and Mexico. Hommersand[209] listed six species in *Corallina, Jania,* and *Amphiroa* from Japan that were also present in Pacific North America south of Point Conception, or had closely related counterparts there.

Table 6
CRUSTOSE CORALLINE SPECIES AROUND HOKKAIDO, JAPAN LISTED ACCORDING TO DISTRIBUTION IN THE THREE TEMPERATURE AREAS DESCRIBED[26]

Subarctic — Eastern Hokkaido

Clathromorphum circumscriptum
Clathromorphum compactum
Lithothamnium "pac-lem"
Lithothamnium glaciale
Leptophytum laeve

Cold Temperature — Northern Hokkaido

Lithothamnium "23"
Hydrolithon sp.

Warm Temperate — Southern Hokkaido

Lithophyllum yessoense
Neogoniolithon sp.
Mesophyllum sp.

COLD NORTHEASTERN PACIFIC

This region, from Alaska to Point Conception, Calif. (35 N. Lat.), is also rich in coralline algae (Tables 5, 7). Twenty-eight species of nonarticulated coralline algae were described for California by Abbott and Hollenberg[1] and a few more may be added from Alaska. Some of the unusual obligate epiphytes described earlier, such as *Clathromorphum parcum* and *Mesophyllum conchatum,* occur here. Studies by Mason[309] and Lebednik[262-265] have provided data on taxonomy, nomenclature, and distribution of nonarticulated species, but a great deal remains to be done on this long coast.

Point Conception has long been recognized as a floristic boundary for a cool-temperate flora in the north and a warm-temperate flora to the south.[209] Three tropical articulated genera, *Amphiroa, Jania,* and *Haliptilon,* are present in southern California, but absent above Point Conception. Above the Point the coralline flora is dominated by species of *Bossiella* and *Calliarthron* with *Corallina, Lithothrix,* and *Serraticardia* present in more spotty fashion (Table 8). *Arthrocardia, Chiharaea,* and *Yamadaea* are rarely encountered and these genera were not recognized as present in California before 1966.[221]

TROPICAL REGIONS

The information available on coralline algae in tropical regions is much less than for the cold water areas described earlier. Scattered information is available in hundreds of floristic papers, but many of them are old and the species determinations cannot be relied on to agree with current concepts. This is particularly true for nonarticulated coralline algae, where painstaking anatomical study is often required for identification. Articulated coralline algae are easier to identify, although here also monographic work is sorely needed for tropical regions. Therefore, instead of analyzing distribution in the six tropical areas individually (Figure 1, Numbers 5 — 10), they will be considered together.

As might be expected, there are numerous coralline algae that play little part in reef-

Table 7
GENERA OF ARTICULATED CORALLINE ALGAE RECORDED FROM SIX AREAS IN THE PACIFIC OCEAN

	1[a]	2[b]	3[c]	4[d]	5[e]	6[f]
Alatocladia	−	+	−	−	−	−
Amphiroa	−	+	−	+	+	+
Arthrocardia	−	−	+	−	−	?
Bossiella	+	+	+	+	−	−
Calliarthron	+	+	+	+	−	−
Cheilosporum	−	+	−	−	−	+
Chiharaea	−	−	+	−	−	−
Corallina	+	+	+	+	+	+
Haliptilon	−	−	−	+	−	+
Jania	−	+	−	+	+	+
Lithothrix	−	−	+	+	−	−
Marginisporum	−	+	−	−	−	−
Metagoniolithon	−	−	−	−	−	+
Serraticardia	+	+	+	−	−	−
Yamadaea	−	+	+	−	−	−

[a] Alaska, including the Aleutian Islands.
[b] Japan.
[c] Monterey, California.
[d] Mexico.
[e] Galapagos Islands.
[f] Southern Australia.

building, but that occur on reefs, flats and lagoons and in other tropical areas. Three coralline types are present

1. Nonarticulated forms growing on hard surfaces.
2. Thin, nonarticulated species growing on noncalcareous marine algae or on marine grasses.
3. Articulated species.

In transects of sloping underwater plateaus in southwestern Curaçao, coralline algae were found to be important components of the flora.[439,471] Many of them were restricted to certain depths, for example, species of *Lithophyllum* and *Porolithon pachydermum* to shallow water, species of *Tenarea, Sporolithon, Hydrolithon,* and *Lithothamnium* to deeper water, sometimes even below 60 m. *Neogoniolithon solubile* (= *N. megacarpum*) and, especially, *Hydrolithon boergensenii* occurred in a wide

Table 8
THE GENERA OF ARTICULATED CORALLINE ALGAE RECORDED FROM CALIFORNIA[1]

GENERA	COMMENTS
Amphiroa	1 sp. below Point Conception
Arthrocardia	1 sp. in northern California and in Monterey
Bossiella	4 spp. prominent throughout California
Calliarthron	2 spp. common throughout California
Corallina	6 spp. throughout state
Haliptilon	1 sp. south of Point Conception
Jania	3 spp. south of Point Conception
Lithothrix	The single species from British Columbia to Baja California
Serraticardia	1 sp. from the north to the Channel Islands
Yamadaea	The only known locality subtidal at Monterey

range of habitats along the transect lines. These kinds of transects, from shores seaward to where algae disappear, need to be made for other tropical areas as well.

Several genera of nonarticulated coralline algae are best represented in tropical areas (Table 2), and these are probably well represented in all tropical regions. Some major papers that provide floristic information are cited in Littler.[284] Gordon et al.[190] described 15 nonarticulated species in 9 genera from Guam and suggested that there were also several undescribed taxa. Recently Bressan[56] published an article on coralline algae in the Mediteranean Sea (Italy), recording 31 species in 12 genera of nonarticulated coralline algae.

In some areas of the world marl or rhodoliths occur in great abundance, dominating sea floor communities as much as several square kilometers in extent. At one time thought to be primarily tropical, it is now realized that such communities made up largely of coralline algae occur in polar areas as well (Table 9).[24]

A recent report by Cabioch[77] revealed the finding of a marl bottom at the Madeira Islands by a French oceanographic expedition in 1966. The material is comprised mostly of freely branching thalli of *Lithothamnium corallioides* and resembles European marl formations except that the associated floristic components differ somewhat. Among the other coralline inhabitants of this marl bank are species of *Fosliella* and *Lithothamnium fruticulosum*.

Table 9
SOME CORALLINE SPECIES FORMING RHODOLITHS AND MARL[24]

Regions	Taxa	Depths
Arctic to subarctic	Lithothamnium lemoineae	Shallow
	L. glaciale	Shallow to deep
	L. tophiforme	Deep
Temperate to subtropical	Phymatolithon calcareum	Shallow to deep
	Lithothamnium corallioides	
	Goniolithon byssoides	
Tropical	Neogoniolithon strictum	Shallow
	Lithophyllum pallescens	
	Hydrolithon reinboldii	Shallow to deep
	Sporolithon timorense	Deep

Among the articulated coralline algae, *Amphiroa* and *Jania* are the genera recorded most often from tropical regions. *Cheilosporum* (e.g., Solomon I.)[449] and *Haliptilon* (e.g., West Indies),[432] (as *Corallina cubensis*) also are reported from tropical areas, but they are not as widespread as the first two. These tropical entities are mostly delicate forms that are epiphytic or that form sand-filled carpets in shallow water. In the Pacific Ocean, parts of the well-described Raroia Atoll were shown to have a distinctive "*Amphiroa* zone"[143] on the reef flat directly behind *Porolithon onkodes* on the ridge. The fronds of *Amphiroa* (*A. annulata*?) formed a turf about 3 cm thick that was a habitat for myriads of small invertebrates and fishes.

Several floristic accounts which accurately depict articulated coralline algae, particularly those prepared by Dawson and his co-workers, are available for the enormously long and varied coastline between Point Conception (35 N. Lat.) and the border between Peru and Chile (20 S. Lat.). The temperate genera, *Calliarthron*, *Bossiella*, and *Lithothrix*, that are present north of Point Conception disappear in the northern part of this coast. The tropical genera, *Amphiroa*, *Haliptilon*, and *Jania*, appear below Point Conception in southern California and extend southward in the tropics. In the south *Bossiella* appears again in the colder waters of Pacific South America. *Calliarthron*, *Serraticardia*, and *Lithothrix*, do not occur in South America.

Near the center of the Pacific coast of the North American continent is California, an area with numerous floristic records revealing that 11 genera of articulated coralline algae are present (Table 8). Because the change in water temperature is basically gradual from north to south, it is difficult to discover relationships that may exist between algal presence and seawater temperature. Abbott and North[2] listed algae that they found to be tolerant of warm waters, such as in southern California and Mexico. The articulated species that could survive periods of temperature above 20°C included *Amphiroa zonata*, *Bossiella orbigniana*, *Calliarthron cheilosporioides*, *Haliptilon gracilis*, *Lithothrix aspergillum*, and several species in *Corallina* and *Jania*. These authors suggested that species tolerating warm waters would be able to occupy costal areas affected by thermal discharges from nuclear power plants.

The Indian Ocean includes the vast area of eastern Africa from the Cape of Good Hope eastward to the Bay of Bengal and also numerous islands (Figure 1). Floristically, there is only scattered information for this area. In subtropical areas, such as the southeastern coast of Africa, the articulated coralline flora is exceedingly rich, with pure stands of *Arthrocardia* occuring when conditions are suitable. In fact, of the six genera present in southeastern Africa, *Arthrocardia* is the most conspicuous, with species of *Amphiroa*, *Cheilosporum*, and *Corallina* also abundant (Table 3). There is a gradual dimunition in frond size when proceeding northward into tropical areas such as Mozambique and Kenya.

Numerous floristic accounts are available for the Mediterranean and Red Seas, but much reliance has been placed on Hamel and Lemoine[197] for the Mediterranean Sea and Papenfuss[347] for the Red Sea. The Mediterranean Sea contains all the species from Atlantic Europe, including the cold-water *Corallina officinalis*. Probably *Haliptilon squamatum* and *Amphiroa* have a center of distribution in the Mediterranean Sea. The Red Sea has a more tropical flora, with only species of *Amphiroa* and *Jania* present.[347]

BELOW 30° SOUTH LATITUDE

This vast area, including regions 11 and 12 (Table 1), brings us back to cold waters again. The coralline algae here are known only in spotty fashion. Major areas are southern Australia, New Zealand, Chile, Argentina, Antarctica, and islands of the Southern Hemisphere. Several publications on coralline algae from Australia and South America have recently appeared, but distributional data like those for the North Atlantic are lacking. Nonetheless, the coralline components of the submarine floras in places like southern Australia and New Zealand are rich in diversity and numbers, and future investigations should prove fruitful.

Among algae restricted to particular localities are the three species of *Metagoniolithon* which are endemic to southwestern and southern Australia to eastern Victoria and the northern shores of Tasmania.[144] New Zealand lacks *Metagoniolithon*, but still contains a rich flora.[472]

Extending southward along the Pacific South American coastline is *Bossiella*, a genus best represented in the western coast of North America, but including two species, *B. orbigniana* and *B. chiloensis*, which extend into Argentina and even into the Atlantic Ocean at least as far north as Buenos Aires.[362] Also recorded from the coast of Argentina are five species of *Corallina*, and two of *Jania*.

SUMMARY

Ideally, the world could be divided into 12 geographic regions that could be analyzed for their coralline distributions, if enough data were available. The paucity of reliable information from great parts of the world makes the possibility of doing this premature.

Above Cape Hatteras in the northwestern Atlantic several epilithic crustose species have been studied by Adey; this region, and the eastern North Atlantic, are the best known floristically. Prominent genera are *Clathromorphum, Leptophytum, Lithophyllum, Lithothamnium,* and *Phymatolithon. Corallina officinalis* is the only articulated species in this region.

Except for *Clathromorphum*, the same crustose genera occur in the British Isles. The heretofore poorly known epiphytic genera, such as *Fosliella*, are now being studied. In addition to *Corallina*, three other genera of articulated coralline algae are present in the eastern North Atlantic: *Jania, Haliptilon,* and *Amphiroa*.

In Japan and associated areas, the coralline flora is rich, and it may be comparable to that in western North America when enough studies can be made. Two articulated genera, *Alatocladia* and *Marginisporum*, are endemic to Japan. Distributional studies around Hokkaido suggest that coralline distribution is regulated by seawater temperatures.

The long coast from Alaska to Point Conception, Calif. contains at least 28 species in 11 genera of nonarticulated coralline algae and 17 species in 8 genera of articulated forms. Among the articulated taxa *Bossiella* and *Calliarthron* are prominent, with *Lithothrix, Serraticardia* and several other genera less common.

Until more phytogeographical studies can be made it is best to treat the six tropical regions (Numbers 5-10 in Table 1) together. Common nonarticulated genera in tropical areas are *Neogoniolithon, Sporolithon, Hydrolithon, Porolithon, Lithophyllum,* and *Tenarea. Amphiroa* and *Jania* are ubiquitous among articulated coralline algae in the tropics, and *Cheilosporum* and *Haliptilon* are sometimes present. Several nonarticulated species live as marl and rhodoliths in tropical areas, and in nontropical areas as well.

Included in the last two regions considered here (Numbers 11 and 12 in Table 1) are southern Australia, New Zealand, southern South America, Antarctica, and islands in the South Atlantic. All three species of *Metagoniolithon* are endemic to southern Australia. There is a great need to study the rich coralline flora in these southern areas.

Chapter 8

GROWTH AND ENVIRONMENT

INTRODUCTION

The presence or absence of coralline algae, as well as characteristics of their growth, are largely influenced by environmental conditions. Scattered among papers on ecology are data on relationships between coralline algae and substrates, seasons, temperature, light, drying, agitation, seawater chemistry, and other organisms. These are areas where much research is needed. First to be considered are spore and substrate interactions.

SPORES AND SUBSTRATES

Except in special cases where spores and sporelings are modified for an epiphytic or parasitic habitat, spores adhere quickly to hard substrates and form small crusts. The rocks that usually comprise these substrates are uneven, although germination has succeeded on smooth glass and plexiglass. Ogata[335-337] plotted the distribution of settled spores of *Amphiroa ephedraea* (probably = *A. beauvoisii*) on artificial surfaces in Japan. He planted the surfaces at various distances from coralline communities and determined the surface textures on which sporelings were most abundant. Spore attachment was greater on surfaces having mounds 0.5 to 0.75 mm high than on smooth or slightly roughened surfaces.

In experiments in the intertidal of Rhode Island Sound, Harlin and Lindbergh[198] found *Corallina officinalis* to compete successfully with species of *Chondrus, Polysiphonia*, and *Ulva* in colonizing artificial surfaces roughened by different-sized particles. In fact, *C. officinalis* showed optimal development and dominated the surfaces roughened by the smallest particles (0.1 and 0.15 mm diam.) whereas the abundant *Chondrus crispus* and *Ulva* predominated on coarser surfaces (particle size 1 to 2 mm diam.). Additionally, on smooth surfaces only encrusting algae plus the crustose bases of *C. officinalis* grew; fronds of the coralline developed only from bases produced on roughened surfaces. On the basis of their observations, they speculated that perhaps " . . . much of the encrusted red coralline algae on our local shores is the basal form of this species that has not been stimulated to develop erect portions."[198] Harlin and Lindbergh[198] also observed that *C. officinalis* appeared only on surfaces made available in the summer. The absence in the fall and winter could have been due to a lack of available spaces, by competition from quick-settling filamentous red and green algae, or by the absence of spores.

In Great Britain, Jones and Moorjani[241] compared tetraspores and their development in two common, articulated coralline algae inhabiting different ecological niches, namely, *Corallina officinalis* and *Jania rubens*. *Corallina officinalis* is widespread, especially in cooler waters (0 to 20°C surface temperature) and is almost invariably epilithic. *J. rubens* occurs in warmer waters (10 to 30°C) and in the British Isles is common only on more temperate shores in the south, and, it is invariably epiphytic.

On evaluating their findings, Jones and Moorjani[241] found that the characteristics of the spores in the two species correlated very well with their ecology. In addition to the results presented in Table 1, they found that at 10°C *J. rubens* failed to develop whereas *C. officinalis* did, albeit slowly. The mucilaginous sheath around the spores of *J. rubens* is probably a mucopolysaccharide. It is possible that a chemical stimulus triggers the attachment of the spore or sporeling of *Jania* when a suitable host is encountered.

Table 1
A COMPARISON OF TETRASPORES
FROM *CORALLINA OFFICINALIS* AND
JANIA RUBENS[241]

Characteristics	C. officinalis	J. rubens
Spore production/12 hr/conc.	25	9
Average spore diameter (μm)	58	75
Mucilage coating around spores	Thin	Thick
Relative sinking rates	Slow	Fast
Time required for attachment (to glass)	2 days	4.5 hr
Relative security of adherence of spore (to glass)	High	Low
Relative tolerance of attached spores to desiccation	Much	Little
Relative time from germination to first evidence of calcification	Short	Long
Relative diameter of crust	Large	Small
Time from germination to first appearance of frond	12 weeks	8 days

SEASONALITY

The seasonality of growth and reproduction of *Corallina officinalis* in the North Atlantic is poorly known. According to Conover,[104] in Massachusetts this species is "dormant" from November to March, begins growing in April, and attains maximum standing crop from May to September. Yet, Lamb and Zimmerman[473] observed new apical growth in September. Sears and Wilce[376] found conceptacles from April through August. In some winters ice-scouring removes considerable numbers of the fronds.[104] In Europe (Denmark), the fronds of this species cease growing in the summer, are sloughed in autumn, and the perennating bases initiate new fronds in late winter.[365] Obviously more information is needed on the seasonal development of this common species in different parts of the world.

Conceptacle formation is probably seasonal in most species in temperate areas. In surveys of numerous plants of *Clathromorphum circumscriptum* and *C. compactum* Adey[4-5] found conceptacle initiation maximal in November to December and mature asexual conceptacles in January to March. Conceptacle production on the Pacific coast of Japan was presented by Chihara in a series of papers.[94-100] Species of *Lithophyllum*,

Dermatolithon, and *Amphiroa* produce reproductive organs beginning in the summer and extending into the fall. Thus reproduction occurs in these taxa at the warmest time of the year. These findings contrasted sharply with those for *Lithothamnium, Marginisporum, Calliarthron, Serraticardia,* and *Corallina* where conceptacles are present throughout the year, but with the fewest in the summer. It is interesting that the summer reproducers are in the Lithophylloideae and the Amphiroideae whereas the others are in the Melobesioideae and the Corallinoideae. Presumably, Chihara's data were for tetrasporangial and bisporangial conceptacles. Data like these are unavailable for other parts of the world and it remains to be seen if correlations exist.

TEMPERATURE

Water temperature plays an important role in determining the forms, growth rates, and taxa of coralline algae in various parts of the world. Generally, calcareous organisms are more abundant in tropical waters than in cold waters. Probably this is due to the fact that tropical waters are supersaturated with calcium carbonate and that saturation decreases in colder waters because of lower temperatures and a higher partial pressure of carbon dioxide. As a result, the amounts of calcite precipitated decreases. However, coralline algae are a prominent exception to this generalization. Consider the massive banks of marl and rhodoliths in subpolar waters and the thick crusts of *Clathromorphum* in the North Pacific. Coralline algae are also common in tropical areas and, in fact, sometimes are more important than coral animals in producing calcium carbonate. Littler[288] stated that the relative dominance of calcareous macroalgae over reef building corals in the western Pacific increases from the equator towards the subtropics and in the west to east direction.

The effects that light and temperature have on growth rates in some melobesioidean species from the North Atlantic was presented by Adey.[12] On the basis of geographical distribution and growth in tanks, he suggested that upper or lower temperature limits determine the distribution of these algae by acting to control vegetative growth or the production of conceptacles. Marginal growth rates of *Clathromorphum circumscriptum* are highly light-dependent at high temperatures but not at low temperatures, indicating that photosynthesis is probably limited at the high temperatures. Because *C. circumscriptum* does not extend south of Cape Cod even when winter temperatures are low enough for reproduction to occur, it may be maximum summer temperatures that are limiting its spread southward, at least in the western North Atlantic.

The possibility that temperatures must go below 2 to 3°C for *C. circumscriptum* to reproduce is a hypothesis that Adey[15] tested. He had previously reported[4-5] that tetrasporangial and sexual conceptacles are rare and that most plants produce bisporangial conceptacles, probably without meiosis. In the Gulf of Maine and elsewhere, conceptacle production begins in September and October, and by January and February they are mature. His experiments on specimens from Maine took place in aquaria under controlled conditions and revealed that conceptacle development took place only when the water temperatures were lowered to 0°C. Even more reproduction occurred when day length was concurrently reduced from 12 to 6 hours per day, an environment which simulated winter where the algae normally grew. In the same paper, he reported that compensation point (O_2 exchange) at low temperatures (0.3°C) was attained at very low light intensities (35 lux), another factor revealing why this species is so successful under arctic conditions.

In summarizing the data on temperature related distributions in the North Atlantic, Adey[14] suggested that there are two major groups: a subarctic group requiring winter temperatures below 1 to 4°C and a boreal group requiring temperatures above that level. The subarctic group includes *C. circumscriptum, Leptophytum laeve, Litho-*

thamnium glaciale, and *Phymatolithon laevigatum*, and the boreal group contains *Lithophyllum orbiculatum*, *Phymatolithon polymorphum*, and *Lithothamnium sonderi*.

It has been shown by several workers that the amounts of magnesium in the carbonate cell walls vary with water temperature.[320] Chave and Wheeler[91] studied *Clathromorphum compactum* by X-ray diffraction analyses of tissues laid down in the summer and in the winter. They were able to do this because the conceptacles are produced in the fall and winter and serve as markers of seasonal growth. In the Gulf of Maine where water temperature varies from 0 to 13°C, *C. compactum* deposits as much as 14 mol% magnesium carbonate in the summer and consistently lower values (as low as 10 mol%) in the winter.

LIGHT

The quantity and quality of the light impinging on coralline algae certainly have a strong influence on their growth and development. Very little is known about this aspect of the environment. There are reports of coralline algae growing in bright sunlit conditions, such as in low intertidal zones and on tropical algal ridges, as well as reports of plants growing under very dim light, such as in deep water, for example, *Jania capillacea* below 60 m in the Eniwetok lagoon.[183]

Littler,[286] in the first published survey of coralline algae made by scuba diving, found that crustose species covered 38.3% of the surface area at 8 to 28 m depths, off Oahu Island, Hawaii. *Hydrolithon* was the dominant genus, with *H. breviclavium* most prominent at sites below 20 m. Surprisingly, *H. reinboldii* was also relatively common. This species is present in the shallow reef flat at Waikiki which is inshore from one of the deepwater sites surveyed. The unusual *Tenarea tessellatum* also grew at this deepwater site.

The prominence of crustose Corallinaceae in temperate deepwater areas was described by Sears and Cooper,[375] who observed them in the Gulf of Maine. The lowermost association was dominated by *Lithothamnium glaciale* and extended from 37 m to the extinction depth at nearly 45 m. In fact, below 40 m, encrusting coralline algae were the only macroscopic algae visible; they covered from 10 to 75% of the available rocky substrate.

An unusual example of the ability of coralline algae to succeed under low light conditions was reported by Dellow and Cassie.[131] In an intertidal cave in New Zealand, the crustose holdfasts of *Corallina officinalis* grew at intensities of only 100 to 500 lux.

The other extreme, that is, the ability to tolerate bright sunlight, seems to be limited to a few species. In southern Australia Shepherd and Womersley[401] noted that *Corallina* spp. and *Amphiroa anceps* withstood full sunlight at low tides. Saito et al.,[475] considered *Corallina pilulifera* in Japan to be a "sun form" because of its tolerance of high light intensities in intertidal areas.

Ridge forming nonarticulated coralline algae, such as species of *Porolithon* and *Lithophyllum*, have been examined for their abilities to grow in bright light. In plants of *L. congestum* from St. Croix growing in high light intensities, the protuberance apices tend to broaden and form slight central depressions. Steneck and Adey[416] suggested that this lateral enlarging may be light activated and result in the anastomosing of adjacent branches when they come in contact. This makes for greater strength as well as increased surface area, both probably of advantage to the plants.

The way *Porolithon onkodes* forms massive crusts in intertidal reef areas would indicate that the plants thrive under stress from increased water motion and insolation. Littler and Doty[289] presented physiological data that reveal its unusual adaptive ability, particularly with regard to bright light intensities. Productivity (mg $C \times 10^{-4}/cm^2/min$)

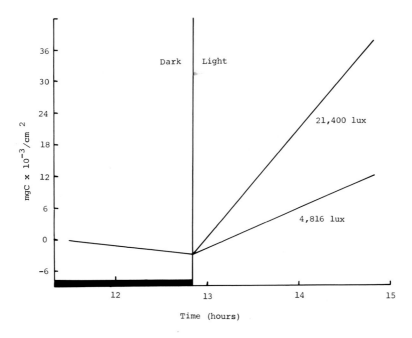

FIGURE 1. The productivity of *Porolithon onkodes* at two light intensities in a strong current. Adapted from Littler and Doty.[289]

was used as an index of response to varying imposed conditions (Figure 1). The experiments were carried out in special plastic containers under laboratory conditions.[285] In summary, the productivity of *P. onkodes* was not limited by water movement, lowered temperatures, such as those caused by evaporative cooling in the intertidal (although productivity was slightly higher at higher temperatures), and high light intensities. In fact, it appears well adapted to bright light; it is light saturated at 21,400 lux. This contrasts with a shade-dwelling coralline species *Sporolithon erythraeum* which is photo-inhibited at intensities this high.[285]

When coralline algae have been cultured under laboratory conditions relatively low light intensities have resulted in the best growth. Intensities below 1000 lux, and even as low as 35 lux for plants from deep water, have been found to be optimal for various species from the northwestern Atlantic.[12,102]

Light intensities influence population ecology, as shown by Adey and Vassar,[29] who installed clean surfaces in tropical areas in the Caribbean. Early successional stages in dimly lit areas contained *Leptoporolithon fragile* and *Tenarea prototypa,* both delicate crustose algae that later gave way to a *Neogoniolithon*-dominated climax community. In deeper water where there is relatively high turbidity, thin, leafy, shade-adapted crustose coralline algea in *Neogoniolithon, Mesophyllum, Sporolithon,* and *Lithophyllum* grow among the fleshy algae, primarily *Sargassum*. These grow slowly and play little role in reef-building.

DESICCATION

Coralline algae in general cannot tolerate desiccation and rarely occur in areas where they are exposed by low tides. Exceptions to this in the eastern Pacific are *Corallina vancouveriensis,* and in the western Pacific *Corallina pilulifera,* species often locally abundant on sloping intertidal rock surfaces. At low tides the densely-branched fronds of *C. vancouveriensis* hang draped on intertidal rocks and retain considerable water among the branches.[231] *Corallina pilulifera* forms moist carpets on intertidal rocks in Japan.

The sensitivity of coralline algae to desiccation was pointed out by Johansen[228] in a study of shorelines uplifted during the 1964 Alaskan earthquake. Strikingly evident, even at a distance of several miles, was the conspicuous white, horizontal belt of bleached calcareous organisms, mostly coralline algae. At certain sites in Prince William Sound, the bleached coralline zone extended more than 10 m above mean lower low water in uplifted areas. These kinds of earthquake-uplifted coralline shores had been seen previously after other Alaskan earthquakes (e.g., in 1899). The uplifted crustose coralline algae were *Clathromorphum circumscriptum* and *Lithothamnium* sp., and the articulated species that were hanging in whitened clusters from the rocks were *Bossiella orbigniana* ssp. *dichotoma* and two species of *Corallina* that were often epiphytized by *Clathromorphum reclinatum*.

Further to the west, at Amchitka in the Aleutian Islands, the Milrow Nuclear test resulted in an uplifted fault of 15 cm. Two species of *Clathromorphum* and *Corallina pilulifera* died from exposure.[61]

The requirement of articulated coralline algae for shade and moisture was shown by Dayton[127] for species of *Corallina, Calliarthron, Serraticardia,* and *Bossiella* in low intertidal zones on the coast of Washington. These algae were placed in his "obligate understory" group and form dense growths under canopies of larger algae or in tide pools where they remain permanently wet. Experiments showed that when the canopies were removed the articulated fronds died and fell off, but the crustose holdfasts remained alive as long as they were not exposed to prolonged desiccation. Under these experimental conditions the plants were reproductively dormant because the fronds and their conceptacles had been lost.[127] The sensitivity to desiccation among these coralline algae related to the fact that most populations appeared to be obligate understory species beneath low intertidal kelp canopies.

WATER MOTION

Coralline algae commonly grow well where water motion is pronounced. The success of *Porolithon onkodes* on algal ridges and *Tenarea tortuosa* in the Mediterranean make these excellent examples of cumaphytes. Working in southern Australia, Shepherd and Womersley[402] found an articulated coralline community at a depth of 7 to 8 m that was restricted to rough water sites. The algae were *Amphiroa anceps, Haliptilon cuvieri* and *Metagoniolithon charoides*. In a later publication,[403] they reported on an almost pure community of *H. cuvieri* that extended upward into the lower intertidal under conditions of much wave action, but that in calm areas was replaced by short tufted plants of *Jania fastigiata*. In other areas similar observations have been made.

Corallina officinalis is particularly abundant, sometimes forming "*Corallina* meadows", in areas along the west and north coast of Iceland.[325] Its presence there and its absence in the east and in much of the south of the island is possibly due to seawater temperatures influenced by currents. Munda[325] observed that differences in morphology are probably related to wave action. In exposed sites the fronds were 1 to 3 cm tall, regularly branched and rigid, resembling *C. officinalis* var. *compacta*. In tide pools the fronds were as much as 12 cm long, less branched, and flaccid.

There are also published examples of the effect of wave action on protuberance development in nonarticulated coralline algae. In *Lithophyllum congestum* at St. Croix, Steneck and Adey[416] found the largest and most complexly branched plants in somewhat sheltered areas and the less-branched specimens on wave-pounded algal ridges.

In addition to plant form and luxuriance of growth, productivity is probably also influenced by water motion. Smith and Kinsey[410] measured calcium carbonate production rates of several Pacific reefs and noted that it is about 4 kg/m²/yr in the shallow seaward portions, but only about 0.8 kg/m²/yr in the protected backreef areas. They interpreted this large difference to be a function of variable water motion.

Living rhodoliths occur on unstable bottoms where there is supposedly enough wave action to periodically turn or agitate the plants. Form provides a clue as to the amount of surge or, more specifically, the frequency and conditions of rhodolith turning. Attempts have been made to classify rhodolith form, in an effort to correlate form and undersea conditions.

To this end distinctions among four rhodolith types were suggested by Bosellini and Ginsburg[51] in a Bermuda study (see Figure 5, Chapter 3)

1. Branching forms, in which thin branches as long as 2 to 3 cm emanate from shapeless central bodies.
2. Globular forms roughly spherical in shape.
3. Columnar forms, in which the rhodoliths bear stocky branches 1 to 2 cm long and 3 to 10 mm wide.
4. Laminar forms, which are made up of layered crenulated crusts 0.5 to 1 mm thick.

Their observations showed that the smooth-surfaced, laminar types were concentrated in open, sandy channels where they frequently rolled. The small, more fragile, branching rhodoliths (= marl) were present in the turtle grass (*Thalassia*) where they probably moved very little. Also immobile were columnar types which often were half buried in sand or rocky depressions.

Given enough time, rhodolith morphology is dynamic, depending on habitat. Bosellini and Ginsburg[51] pieced together morphological series and suggested that branching types may develop into laminar types via the columnar type. As the structures are agitated, the growing branch apices tend to flatten and expand (Figure 2). Eventually contiguous branch tips anastomose and form a smooth-surfaced rhodolith of the laminar type. The history of development can be construed in sectioned samples.

Subsequently, Bosence[52] explained further the transformation of open branching rhodoliths to densely branched forms (the latter possibly equal to the globular or columnar types),[51] by wave action. In a study of *Lithothamnium corallioides* and *Phymatolithon calcareum* marl in a bay in western Ireland, he described the process by which greater rolling frequency results in denser forms due to the increased abrasion and damage to the branch apices. Apical growth is thus arrested and, if rolling continues, growth will be restricted to undamaged cortical cells below the branch apices (Figure 2). Instead of elongation, there will be a broadening of the branches by meristematic activity located within the protection of adjacent branches. This will lead to thicker branches and less space between the branches. Continued growth would result in the anastomosing of adjacent branches, as described by Bosellini and Ginsburg[51] in deriving their laminar type of rhodolith.

In order to understand the effect of waves on the frequency of movement, Bosellini and Ginsburg[51] painted red the upper surfaces of 50 rhodoliths in shallow Whalebone Bay, Bermuda. These were replaced and observed to be rolled about at least weekly in the winter and less often in summer. Wind velocities of 15 knots and waves 60 cm high resulted in movements of several meters even in specimens as large as 8 cm in diameter.

The energy of water movement decreases as depth increases and problems have been encountered in attempting to explain a periodic rolling of rhodoliths living in deep

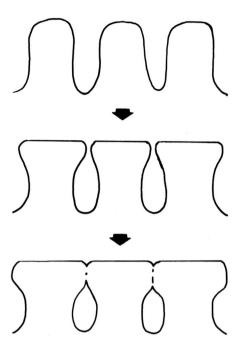

FIGURE 2. Stages in the anastomosing of protuberances of marl of *Lithothamnium corallioides*. The continued lateral growth and the suppression of apical growth results in the fusion of contiguous protuberances. It is probably due to abrasion caused by rolling. Adapted from Bosence.[52]

water. There are three unresolved hypotheses for explaining the formation of rhodoliths in deep water.[51]

1. They are moved by storm waves and internal waves.
2. They no longer move, having developed during glacial periods when water levels were lower and surface waves provided the energy.
3. The rhodoliths are moved by a shifting of the sand in which they occur, less energy being needed for the movement of sand grains.

Ripple marks have been noted in deep water, and McMaster and Conover[314] showed that at nearly 100 m in the Canary Islands rhodoliths were lined up in troughs in the sand. Adey and MacIntyre[24] mentioned that rhodoliths probably form at least to depths of 60 to 70 m. The question of the forces moving deep water rhodoliths cannot be resolved until further research is carried out.

SEAWATER INGREDIENTS

The lack of research on coralline algae growing in defined media allows only for a consideration of some observations on their growth related to salinity, pollution, and phosphate. On the basis of field observations it is apparent that most coralline algae cannot grow where salinity is less than that of the open ocean. Doty and Newhouse[476] emphasized the stenohaline nature of *C. officinalis* in New Hampshire as did Conover[104] in southern Cape Cod (near Woods Hole) who found it only at the entrance (salinity above 30 0/00) of the estuary he studied. Possibly more than any other species, *Phymatolithon polymorphum* tolerates low salinity levels and even grows in water at 13 0/00 in Norway.[14]

Conflicting reports have been made of the effects that domestic sewage discharged into the ocean in southern California has on articulated coralline algae. The first reports were by Dawson[115,121] who observed that *Bossiella insularis* (= *B. chiloensis*), *B. cooperi* (= *B. orbigniana* ssp. *orbigniana*), *Corallina vancouveriensis*, and *Lithothrix aspergillum* formed nearly pure intertidal communities near sewage outfalls, whereas they were not especially abundant in nearby cleaner areas. Studies at San Clemente Island to precisely compare intertidal communities receiving untreated domestic sewage and comparable ones not receiving sewage were reported by Littler and Murray.[291] They looked for any possible effects that the sewage would have on the coralline algae and found that *Corallina officinalis* var. *chilensis* was equally as abundant at the outfall as in the control areas. Other species that were present at the control sites, namely, *C. vancouveriensis* and *L. aspergillum*, as well as some crustose species, were notably absent from the outfall area. The discrepancies between Dawson's observations and Littler and Murray's findings suggest that experimental work is needed and that the effects may vary according to taxon.

This led Kindig[247] to carry out a quantitatively oriented study on *Corallina officinalis* var. *chilensis*, *Corallina vancouveriensis*, *Bossiella orbigniana*, *Haliptilon gracile*, *Amphiroa zonata*, and *Lithothrix aspergillum* in southern California. He studied growth and productivity in polluted areas and cultured plants in the laboratory in sewage effluent. In culture studies two months in duration, *A. zonata*, *B. orbigniana*, and *C. officinalis* var. *chilensis* grew in primary effluent at a rate equal to or faster than did the controls. Different populations of *C. officinalis* var. *chilensis* reacted differently to pollution. This differential response in a single species suggests physiological acclimatisation or genetic adaptation to the stresses of living in sewage effluent and would signify caution in using benthic algae as biological indicators of pollution. Two species, *C. vancouveriensis* and *H. gracile*, grew less well and exhibited depressed net productivity in primary effluent. Kindig's[247] results seem to agree with Dawson's observations with respect to net productivity rates for *B. orbigniana*, *C. officinalis*, var. *chilensis* and *L. aspergillum*. These more tolerant species showed lower net photosynthetic quotient (PQ) values than did those species that were relatively intolerant. It may be that chelating agents determine the growth and productivity responses of coralline algae or that these factors influence calcium and magnesium incorporation into the growing cell walls.

The apparent enhancement of growth in sewage-enriched seawater seems to be at odds with the finding that some articulated coralline algae are intolerant of phosphate concentrations that generally stimulate growth in other marine algae.[419,420] *Jania rubens* grew better at one fourth the usual phosphate concentration in culture media (von Stosch's Grund Medium) as did *Corallina officinalis* and *Cheilosporum sagittatum*, and a species of *Jania* from southern Australia.[58] *Corallina officinalis* from Europe, cultured by Von Stosch[420] was not sensitive to phosphate, indicating differential reactions by ecological races of this species. Possibly phosphate ions interfere with crystallization, if the concentrations are too high. Obviously, more work is needed on phosphate sensitivity of coralline algae.

BIOTIC INTERACTIONS

The shallow water habitats that coralline algae occupy are also suitable for a great variety of other kinds of organisms, some of which have evolved adaptations allowing them to more successfully cohabitate with coralline algae. The length of time that

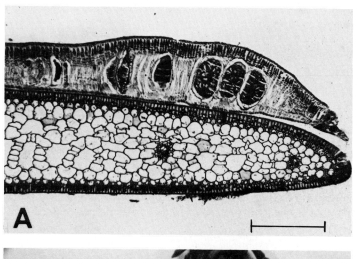

FIGURE 3. Epiphytic coralline algae. A. Section of *Melobesia pacifica* growing on a leaf of *Phyllospadix* (angiosperm). Scale = 300 μm. Courtesy T. Masaki. B. *Jania pusilla* on a branch of *Cystophora* (Phaeophyta). Scale in millimeters.

coralline algae have lived in the seas has resulted in adaptations for living in association with other marine plants and animals. The relationships may be loose, such as animals living among the branches of articulated coralline algae, or tight, such as the obligate parasites in coralline algae. They may be neutral and have little effect on the coralline algae, they may be negative and result in harm, or they may be positive, such as when the plants are provided with a suitable substrate on another plant (Figures 3 and 4). The coralline algae that are apparently obligate epiphytes or parasites and that are visibly modified for their habitat were treated earlier (Chapter 3).

The dense fronds of articulated coralline algae often serve as habitats for numerous small animals. Dommasnes[137,138] and Hagerman[196] reported on the diverse fauna on *Corallina officinalis* in western Norway, with the first emphasizing animal fluctuations and the second ostracod populations. Species of *Corallina*, *Cheilosporum*, and *Jania* also support many animals in New Zealand.[207] Very little is known about the possible adaptations of animals inhabiting coralline fronds.

In experiments with *Clathromorphum circumscriptum* Adey[15] found that for successful growth in aquaria limpets needed to be included. When the animals were ex-

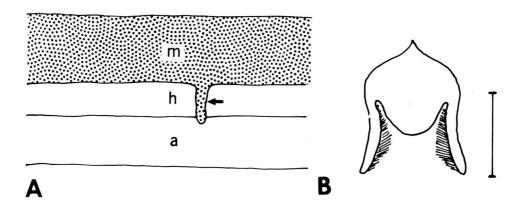

FIGURE 4. Unusual organismal associations. A. Diagram showing manner in which *Metagoniolithon stelliferum* (m) grows on *Heteroderma cymodoceae* (h) which in turn grows on the marine grass *Amphibolus antarctica* (a). The crustose holdfast of *M. stelliferum* has produced an outgrowth (arrow) through *H. cymodoceae* into the outer part of the grass leaf. Adapted from Ducker and Knox.[146] B. Fruits of *Phyllospadix* which become attached to articulated coralline algae, such as *Calliarthron tuberculosum* off central California. Scale = 5 mm. Adapted from Gibbs.[182]

cluded the epithallia were 2 to 4 cells thick, lacked the usual plastids, and the surface was covered with layers of blue-green algae. However, when limpets were present the epithallia were 4 to 5 cells thick, the usual plastids were present, and the blue-green algae mostly absent. Possibly the animals kept the coralline algae free of epihytes and also, as Adey[15] suggested, the plants require the periodic removal of epithallial cells for best growth. In most coralline algae epithallial tissues are apparently naturally sloughed, but perhaps this ability has been lost in *C. circumscripticum* and browsing animals are required to perform this function and stimulate healthy growth of the plants.

Examinations of gut contents of invertebrate animals have often revealed fragments of coralline algae, for example, *Jania capillacea* in species of the gastropod *Turbo* in the South Pacific.[436] It is doubtful that much if any nourishment is derived from such food. Paine and Vadas[344] obtained calorific values for numerous benthic algae in Washington State and found these values to be very low, as did also Littler and Murray[477] for *Haliptilon gracile*, *Corallina officinalis* var. *chilensis*, and *Lithothrix aspergillum* at San Clemente Island off southern California. In southern California the Black Abalone *Haliotis cracherodii* Leach lost weight when restricted to a diet of coralline algae, whereas the animals grew when fed most other algae.[268] Paine and Vadas[344] and Littler and Murray[477] speculated that calcification and low calorific values are of survival importance under high grazing intensity.

As will be described later, fish ingest coralline algae and, in several ways, play an important role in tropical submarine ecology. Fragments of coralline algae, for example, articulated forms in *Jania* and *Amphiroa*, have been found inside fishes.[478]

Several taxa of ascomycetous fungi grow on coralline algae. Kohlmeyer[249-250] described *Mycophycophila corallinarum* (Crouan et Crouan) Kohlmeyer from *Corallina elongata* (as *C. mediterranea*) and other coralline algae. This fungus grows on various algal hosts, but the close relationship between the perithecia and epiphytic coralline algae (e.g., *Melobesia* and *Dermatolithon*) induced Kohlmeyer[250] to suggest that a symbiotic relationship exists between *M. corallinarum* and the epiphytic algae. They may form a " . . . primitive lichen in which at least one partner, the alga, can live independently from the other". Bonar[44] described another species of *Mycophycophila*, *M. polyporolithi* Bonar, from the lower surfaces of *Mesophyllum conchatum* (as *Polyporolithon conchatum*). Kohlmeyer[249-250] has also reported *Lulworthia kniepii* Kohlmeyer from species of *Lithophyllum* in the Mediterranean Sea and the Canary Islands. All of these ascomycetes are prolific producers of perithecia.

An unusual condition of one epiphyte growing on another has recently been reported from southern Australia by Ducker et al.,[145] Ducker and Knox[146] and Ducker.[144] Of the three species of *Metagoniolithon*, *M. stelliferum*, and *M. chara* are epiphytic, usually growing on the marine angiosperms *Amphibolus antarctica* (Labill.) Sonder et Aschers. ex Ascherson and *A. griffithii* (Black) Den Hartog. The articulated species grow not directly on the seagrass, but rather on nonarticulated seagrass epiphytes in *Heteroderma*, especially *H. cymodoceae* (Figure 4). This unusual situation of what may be called compound epiphytism has been seen numerous times by Ducker and co-workers. The germination of spores of *Metagoniolithon* has succeeded only to the stage of small crustose sporelings. No evidence for the production of rhizoids in germinating spores, such as reported by Cabioch[69,76] for *Amphiroa*, were observed in laboratory cultures of *M. stelliferum*.

Another south Australian epiphyte grows on the common subtidal fucoid *Cystophora: Jania pusilla*.[147] These plants are less than 1 cm long with up to 12 fronds growing from small pulvinate crusts less than 1 mm in diameter (Figure 3). In spite of the small frond size the intergenicula are robust for *Jania* (200 to 500 µm broad) and, among the coralline algae, the tetrasporangia are the largest recorded: 200 to 235 µm long and 100 to 130 µm in diameter. It is interesting that *J. pusilla* resembles the diminutive *J. radiata* from Japan, a species that is also an obligate epiphyte on a fucoid, namely *Sargassum*. In addition to these examples, other species of *Jania* also exhibit host specificity. Duerden and Jones[149] reported that *J. rubens* in the British Isles grows mainly on two species of *Cladostephus* (Phaeophyceae, Sphacelariales).

Two delicate and filamentous red algae are apparently restricted to genicula in several species of articulated coralline algae in western North America: *Acrochaetium amphiroae* (Drew) Papenfuss (Nemaliales) and *Ptilothamnionopsis lejolisea* (Farl.) Dixon (Ceramiales). According to Dawson,[479] they produce rhizoids which penetrate between the outermost genicular cells and establish attachment organs.

The unusual discovery of fruits of the surfgrass *Phyllospadix* attached to branches of *Calliarthron* and *Bossiella* led Gibbs[182] to search for the significance of this relationship. The results were based on specimens from south (Monterey Bay) and north (Bodega Bay) of San Francisco and included fruits in various stages of germination. These fruits and seedlings are attached by two processes with backward-oriented bristles which Gibbs described[182] as resembling a "beetle's head with rigid, bristle-fringed antennae" (Figure 4). These become attached in surf zones to the coralline algae where they germinate, the roots growing downward and eventually becoming established in the sand or rock below.

There are few examples of a direct relationship between coralline algae and *Homo sapiens*, except the occasional shipwreck on an algae-produced reef. The use of marl as a soil conditioner has been practiced in England since the 18th century and in France since the 19th century.[40] The annual harvest of more than 300,000 tons of marl, comprised of *Phymatolithon calcareum* and *Lithothamnium corallioides*, is presently carried out off Brittany where deposits of hundreds of millions of tons are estimated to occur. Most of the commercially collected marl is dead and greyish to brownish white. The algae are scooped up by grabs and dumped into the holds of ships holding 200 to 500 tons. After harvest the marl is trucked to factories, dried in large rotary driers, reduced to the desired particle size and bagged for shipping. According to Blunden et al.,[40] the granular to powdery marl is used with success in most west European countries as a soil conditioner. The effective rate of activity varies with particle size, with a typical plot of farmland receiving 400 kg/acre every 3 years. Marl is also used as a cattle and pig food additive and in the filtering of acidic drinking water. This removes toxic metals such as lead, copper, and zinc by absorption, by ion-exchange, and by precipitation in the alkaline environment.

Apparently the only ingestion of coralline algae was when it was used as a vermifuge. Preparations of *Corallina officinalis* and *Jania rubens* were used in Europe for this purpose until the end of the 18th century.[480-482]

SUMMARY

Numerous factors in the environment influence whether or not coralline algae will be present in a particular locality and how well they will grow there. These factors are substrates, seasons, temperature, light, drying, agitation, seawater chemistry, and other organisms.

Most coralline spores germinate best on hard rough surfaces. The spore and germination characteristics of the epilithic *Corallina officinalis* and the epiphytic *Jania rubens* from the British Isles differ in several respects, such as size, amount of investing mucilage and rate of germination.

As a group coralline algae occur in cold as well as tropical waters, but the included taxa are usually restricted to certain regions, probably a factor of water temperature. *Clathromorphum* is most prevalent in arctic waters, something which may be regulated in the various species by temperature control of reproduction or vegetative growth. Other taxa may be under similar control. Magnesium is incorporated in the carbonate fraction of cell walls more in summer than in winter.

Most coralline algae grow best in dim light; a few grow best in bright light. Light intensity probably plays a role in the depth distributions of various species. *Porolithon onkodes* is among those few species that live in bright light.

Most coralline algae will die if brought out of water for any length of time, such as when earthquakes in Alaska elevate the shores. A few species seem able to tolerate desiccation better than most species.

Many species seem to require considerable water motion for best growth, but very little is known of this. Morphology is apparently affected by agitation, in crustose species with protuberances, articulated fronds, and unattached forms.

Very few species are able to tolerate lowered salinity, another factor which seems to influence where coralline algae grow. Domestic sewage in southern California has a positive effect on the growth of some species of articulated coralline algae, and a negative effect on other species. An unusual inability to tolerate concentrations of phosphate that are usually required by noncorallinaceous plants has been found in some articulated coralline algae.

Many relationships exist between coralline algae and other organisms. Some are only loose cohabitation with animals and some involve ingestion. *Clathromorphum circumscriptum* may even benefit from having epithallial tissue removed by limpets. Fungi grow on some coralline algae and epiphytism by or on noncalcareous algae is common. Several species of coralline algae live on various marine plants (e.g., angiosperms and fucoid brown algae). Man gathers large amounts of marl off Brittany and uses it as a soil conditioner.

Chapter 9

PRODUCTION

INTRODUCTION

Although coralline algae are notoriously slow growing plants, the fact that they are extremely abundant, often grow in dense stands, and are widespread in warm and cold waters, makes it important to consider their production. It is also prudent to emphasize that there are two major kinds of products, namely, organic and inorganic material. It is the production of the latter in the form of calcium carbonate that makes coralline algae so unique among marine plants. As mentioned earlier, other benthic marine algae produce calcium carbonate, but none of them produce such concentrated amounts as do coralline algae. Thus it behooves us to consider the ecological and community aspects as they relate to organic and inorganic production. Also included in this chapter are growth rate studies as well as establishment and successional aspects of communities dominated by coralline algae. Finally, the immediate fate of the plant remains, and the long-range products of recrystallization and sediment formation, are considered.

GROWTH RATES

Growth is usually determined by measuring rates of marginal extension and increasing thickness in crustose coralline algae, and increases in branch length in articulated coralline algae. Other measurements indicating growth are sometimes given, such as dry weight and organic and carbonate productivity.

Growth rates are generally greatest in warm waters. Adey and Vassar[29] determined marginal extension rates in shallow water tropical crustose species at 1 to 2.3 mm/month and pointed out that this is about twice as fast as for those in boreal and subarctic waters. The maximum accretion rates (meaning increase in thickness) of about 5 mm/yr are much greater than for cold water species. For *Lithophyllum congestum* at St. Croix, Steneck and Adey[416] found protuberance elongation of up to 8 mm/yr, increases in crust thickness up to 5.2 mm/yr, and marginal extensions of at least 10.8 mm/yr.

Few estimations of rhodolith growth rates have been made. Bosellini and Ginsburg[51] examined rhodoliths on Bermuda shipwrecks (wrecked 1943) and estimated growth rates (probably increased thickness) in columnar forms to be 0.4 mm/yr. In their study, average-sized rhodoliths had outer coralline coatings 3 cm thick which, without significant interruptions, would have taken 75 years to form. Adey and MacIntyre[24] suggested that continuously growing deep water rhodoliths 20 to 30 cm in diameter might be 500-to-800-years-old. Periodic interference with growth by occasional burial, exhumation, and damage by browsing animals would extend that estimate.

Several reports of growth rates in articulated coralline algae have also been published. Off central California a kelp bed population of the relatively robust *Calliarthron tuberculosum* had branches that elongated 1.7 mm/month with the fastest growth from October to February, a time of maximal light penetration through the kelp bed.[234] In *C. tuberculosum* the medullary cells are organized in tiers 50 to 75 μm tall and the intergenicula, which are 2.5 to 4.5 mm long, contain 40 to 72 tiers (Figure 1). Calculations showed that 27 cellular tiers are produced per month, or about one a day. Every 45 to 80 days a genicular tier of cells is produced. Plants germinated *in situ* from spores produced conceptacles in one year and fully developed fronds 10 cm

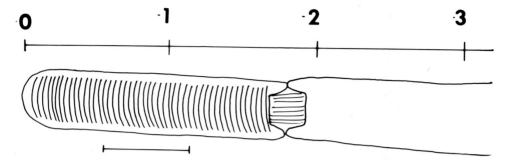

FIGURE 1. Outline of branch tip of *Calliarthron tuberculosum* from a depth of 11 m in a kelp bed off central California. Increments of growth are based on an elongation rate of 1.7 mm/month. The months are numbered at the top. Arching lines indicate the transverse cell walls of medullary cells. Scale = 1 mm. Adapted from Johansen and Austin.[234]

tall in four to five years. In a study of succession in a *Macrocystis* bed off Santa Cruz Island in southern California, Foster[167] included *Calliarthron cheilosporioides* and *Corallina officinalis* var. *chilensis* among the slow growing perennial algae. The fronds elongate between 1.1 and 2.2 cm/yr (the measurements were probably made on *C. cheilosporioides*).

Corallina officinalis from Massachusetts, with fronds that are more delicate than in *Calliarthron*, grew 2.2 mm/month in culture.[102] Based on that rate an intergeniculum about 1 mm long was produced every 12 days and, assuming that the tiers of medullary cells are 54 μm high, 1.5 tiers (equal to 80 μm) formed each day. These rates for *C. officinalis* are slightly greater than for *Calliarthron tuberculosum* in the field. Lowering the temperature and decreasing the light intensity lowered the elongation rate in *C. officinalis*.

COLONIZATION AND SUCCESSION

Epilithic coralline algae grow in nearly pure stands in shallow waters in many parts of the world. Such populations may be considered coralline-dominated climax communities although, if clean surfaces are available, the successional stages may be very brief as in *Calliarthron cheilosporioides* communities in kelp beds off the coast of southern California.[167] Fresh Hawaiian lava flows (in 1955) were covered by fertile fronds of *Corallina sandwichensis* the second year after the lava entered the sea.[140] This repopulation of denuded areas by coralline algae in the first or second year has also been seen in Great Britain[248] and elsewhere.

In a successional study of marine algae colonizing vertical surfaces of newly installed breakwaters in southern Hokkaido, Japan, Saito et al.[367] showed that *Corallina pilulifera* and noncorallinaceous perennial algae became increasingly prominent until climax formations were attained in three to four years. Furthermore, as more plants of *C. pilulifera* became established they occupied higher levels. On one-year-old surfaces the populations of *C. pilulifera* extended up to 10 cm above mean low water. On three-year-old surfaces they extended to 90 cm, and on breakwaters built 16 years previously, they were at the 110 cm level. On a percent coverage basis *C. pilulifera*, the only coralline included in this study, exceeded other perennial algae on substrates more than 16-years-old. Working in Yoshimi, Yamaguchi Prefecture, Katada and Matsui[244-245] also studied community restoration in denuded areas that had been dominated by *C. pilulifera* and concluded that this species would comprise the climax community, although succession passed first through crustose and erect, annual noncorallinaceous species.

The disturbance of an underwater community and a subsequent colonization by articulated coralline algae was described for a shallow reef off southern California.[352] Here the reef was dominated by mussels (*Mytilis edulis*) which were almost eliminated in one year, probably by an invasion of the seastar *Pisaster giganteus* (Stimpson). The coralline replacement was *Corallina officinalis* var. *chilensis*, which soon formed a uniform mat 6 cm thick. Pequegnat[352] suggested that the uniform populations of *C. officinalis* var. *chilensis* that he observed in several places in southern California are associated with disturbed conditions and that their growth is enhanced by human pollution.

Kelp forests dominated by *Macrocystis pyrifera* (L.) C. Ag. form extensive communities off the western coast of North America. In many of these forests dense growth of shade-tolerant plants constitute an understory that often consists mostly of coralline algae. In a forest at Santa Cruz Island Foster[167] delimited four strata, with crustose coralline algae prominent in the lowermost layer and articulated coralline algae prominent in the layer just above this. Several articulated species in *Calliarthron*, *Corallina*, and *Bossiella* are most common in these kelp forests, although *Lithothrix*, *Haliptilon*, and *Serraticardia* are also present. *Calliarthron cheilosporioides* and *Corallina officinalis* var. *chilensis* are by far the most common algae on the floor of the Santa Cruz Island forest.[167] These species, together with *Bossiella orbigniana*, cover 43% of the total floor of the forest. A comparable situation has been recorded in a *M. pyrifera* forest near Monterey in central California, but here *Calliarthron tuberculosum* was the most common plant and *Bossiella californica* ssp. *schmittii* was second.[234]

ORGANIC PRODUCTIVITY

Tropical reefs are among the most productive systems known. It must be remembered that the products are of two types, namely, organic and mineral. The measurements which have been made on the organic productivity of tropical coralline algae were reviewed by Lewis.[279] Marsh[302] measured the productivity of the reef building coralline algae at Eniwetok Atoll and obtained low values, a gross primary production of 550 gC/m^2/yr and a net production of 240 gC/m^2/yr. Thus, he concluded that these algae are more important in reef building than in primary production. Using several methods ($^{14}CO_2$ and pH electrode), Littler[287] studied the coralline productivity on Hawaiian reefs (Waikiki) and obtained somewhat higher numbers for the five species he studied: gross productivity of 219 to 1241 gC/m^2/yr and net productivity of 183 to 949 gC/m^2/yr. In their work on Hawaiian reefs, Littler and Doty[289] found the net productivity of *Porolithon onkodes* to be 803 gC/m^2/yr and *P. gardineri* 876 gC/m^2/yr. These are similar to measurements by Wanders[441] who obtained production values of 890 gC/m^2/yr gross productivity and 370 gC/m^2/yr net production in a shallow reef flat at Curaçao (Netherlands Antilles). Further data were provided by Sournia[413] for a fringing reef community in French Polynesia dominated by *Neogoniolithon frutescens*. Although gross production for the entire reef community was in the order of 2682 gC/m^2/yr, the gross productivity of the coralline community was only 1387 gC/m^2/yr. On the basis of these data, Lewis[279] concluded that coralline algae have rates of production that are considerably lower than those for reefs as a whole.

Wanders[441] made oxygen production and consumption measurements using the light-and-dark-bottle technique and, based on data from experiments run in the lagoon and in an outdoor tank, calculated light intensity effects. Net productivity becomes positive (i.e., compensation intensity is reached) in *Porolithon pachydermum* at 2500 lux and in species of *Lithophyllum* at 900 lux. As the light intensity increases, these coralline algae reach a peak of productivity (saturation) at 7000 to 8000 lux and indications are that at extreme intensities, such as 70,000 lux, photo-inhibition occurs.

Table 1
PRODUCTIVITY AS SHOWN BY OXYGEN EXCHANGE ($gO_2/m^2/24h$) IN CRUSTOSE CORALLINE ALGAE AND OTHER COMPONENTS OF A SHALLOW REEF FLAT AT CURAÇAO, NETHERLANDS ANTILLES[441]

Taxa or communities	Consumption	Gross productivity	Net productivity
Neogoniolithon solubile (= *N. megacarpum*?)	1.8	4.2	1.4
Porolithon pachydermum	2.0	3.1	1.1
Lithophyllum intermedium	1.8	2.7	0.9
L. daedaleum (= *L. congestum*?)	1.6	2.6	1.0
Lithophyllum sp.	1.6	2.7	1.1
Lithophyllum sp.	1.6	2.3	0.7
Noncalcareous algae	2.2	7.4	5.2
Corals	9.0	14.3	5.3

Among the coralline algae, gross and net productivity was highest in *Neogoniolithon solubile* (= *N. megacarpum* ?) (Table 1). However, the coralline productivity rates were below that for communities of fleshy algae and corals. Of the 2330 $gC/m^2/yr$ net productivity, about one third is produced by noncalcareous algae and by crustose coralline algae and two thirds by coral animals.

The cover and organic productivity of numerous noncalcareous algae as well as three articulated species were determined by Littler and Murray[290] in the rich intertidal community of San Clemente Island off southern California. *Corallina officinalis* var. *chilensis* was among the most important species in cover (second at 14.1%), but somewhat less important in productivity (fifth at 9.4 mg C fixed/h), with *Lithothrix aspergillum* and *Corallina vancoveriensis* relatively low. It appears that the organic productivity of articulated coralline algae is also relatively low.

INORGANIC PRODUCTION

The ratio of the inorganic to the organic component in coralline thalli is much greater than that for noncorallinaceous seaweeds. It is the calcium carbonate fraction of the plants that is most important to a community in which coralline algae are prominent. This component plays a large role in reef building and sediment production in shallow waters. All types of coralline algae, including crusts, articulated species, and unattached forms, are important in inorganic productivity.

The chemical aspects of calcification have been given earlier and we shall now focus on reef growth in tropical areas. Sea levels oscillated rapidly during much of the Quaternary and, as can be imagined, reef growth responded. The recent rising sea has probably placed an upper limit on reef growth rates, and measurements by Smith and Kinsey[410] indicate that the estimated maximum vertical accretion of 3 to 5 mm/yr in algal ridges can be realized only in a rising sea.

Important information on calcium carbonate production in reef growth has been given in a paper by Smith[409] on Eniwetok Atoll. He measured alkalinity depletion in water flowing over reef flat communities and then calculated carbon dioxide dynamics, organic production, respiration, and calcification. Some of the information presented may be summarized as follows:

1. Calcification rates vary little in communities consisting of diverse kinds of calcifiers, whether they are coral animals, nonarticulated coralline algae, or a *Jania* turf that was common on the reef flat.

2. Only an algal veneer a few centimeters thick and scattered coral animals have accumulated on the Eniwetok reef flat in the last 4000 years.
3. Annual mean sea level at Eniwetok oscillated as much as 10 cm/yr between 1952 and 1970, but there was no significant net change in sea level during that period.
4. Sea level before about 7000 years Before Present (BP) probably rose more than 10 mm/yr.
5. Calcification rates of 4000 g $CaCO_3/m^2/yr$ were obtained, and less than 1000 g $CaCO_3/m^2/yr$ were removed by burrowing and grazing animals.

These data presented some anomalous situations. An amount of 4000 g $CaCO_3/m^2/yr$ is sufficient for 3 mm/yr of upward reef growth, and yet there has not recently been a significant change in sea level, and the reef flat has no grown much for the last 4000 years.[409] Smith suggested that the probable sink for this excess calcium carbonate is the Eniwetok lagoon, where sediments accumulate, particularly during storms. Second, an unanswerable question is how the reef could have grown fast enough to keep up with the rapidly rising sea before 7000 years ago. Third, the rates obtained by Smith for calcium carbonate accretion are markedly below those obtained by Odum and Odum[334] for studies in the same area (3×10^4 g $CaCO_3/m^2/yr$). The Odum and Odum figure has been revised downward by other workers and now is closer to Smith's value. Finally, this Eniwetok work showed that calcification rates are about the same at night as in the day and does not agree with the data given by Goreau[191] for which the rates were higher in the day. Smith answered this by suggesting that his measurements were made in flowing water in which a respiration produced buildup of carbon dioxide was not present, such as might have been the case in Goreau's containers.

A Caribbean fringing reef on the west coast of Barbados (Northern Bellairs Reef) has been studied by Stearn et al.[415] with the primary goal of determining calcium carbonate accretion rates. They showed that crustose coralline algae provide most of the reef cover, even to a depth of six meters on the seaward front. In their underwater surveys, they estimated that of the coralline cover *Porolithon* accounted for 40%, *Neogoniolithon* 25%, *Lithophyllum* 20%, and *Mesophyllum* 15%. In their study they used artificial substrates and staining with Alizarin Red-S, arriving at a calculated overall productivity for the reef of 47×10^6 g/yr or 47 metric tons/yr.

The technique of staining living coralline algae with Alizarin Red-S was used in the same manner that has been used to measure coral growth and seasonality for several years. Stearn et al.[415] covered fist-sized areas of *Porolithon* in the morning with large plastic bags in each of which was included about 1 g of the dye. In the late afternoon the bags were removed. Only partial success was achieved when the algae were sacrificed and sectioned three months later. In some of the plants the dye had been incorporated, and distances from the dye line to the thallus surface were measured and found to be 0.5 mm. Assuming that growth was constant throughout the year, the vertical growth rate of 2 mm/yr falls within the finding by Adey and Vassar[29] for growth rates in reef areas of St. Croix. Next Stearn et al.[415] calculated that these coralline algae contained 57.4% calcium carbonate for a bulk density of 1.56 g/cc. Although few measurements were made, the data suggested that accretion was 3120 g $CaCO_3/m^2/yr$. The carbonate productivity that could be ascribed to *Porolithon* on the reef was calculated to be 28.6×10^6 g $CaCO_3/m^2/yr$.

UNATTACHED CORALLINE ALGAE

Branched fragments or roughly rounded nodules of nonarticulated coralline algae occur free-living on many continental shelves from sea level to depths of 200 m.[24]

Publications dealing with ocean floor sediments made up of dead or living thalli or fragments of nonarticulated coralline algae are accumulating rapidly, especially in geological journals. The following summary dealing with biological aspects of living corallines or the derivation of nonliving sediments relies especially on papers by Adey and MacIntyre,[24] Milliman,[321] and Bosellini and Ginsbury.[51] The term "sediments" may be misleading; it refers to particulate matter from the size of sand grains to nodules up to 30 cm in diameter.

As pointed out by Adey and MacIntyre,[24] the terms *Lithothamnion* ball, algal ball, oncolite, algal nodule, rhodolite, rhodoid, marl, and rhodolith have all been used with respect to calcareous particulate matter derived from or composed of coralline algae. With the aid of definitions given in Milliman,[321] two growth forms will be recognized:

1. *Marl* (maerl) — a sand or gravel in which more than half of the sediment consists of dead or living coralline branches or parts thereof. These branches correspond to the protuberances of attached crustose coralline algae (Figure 2, Chapter 8).
2. *Rhodoliths* — nodules or concretions in which coralline crusts have grown around a nucleus. If the crusts are branching forms, they may fragment and provide branches for marl.[52] Sizes range from only a few millimeters to, more often, several centimeters in diameter. Shapes depend on environment and species composition. This will be discussed later.

The distinctions between the more delicate branching structures making up marl and the more massive rhodoliths may blur and become meaningless. In fact, as shown by Bosellini and Ginsburg[51] and by Bosence,[52] appropriate environmental conditions and time may result in a transformation of marl into rhodoliths.

Sediments may also be derived by attrition directly from nearby encrusting coralline algae. The particles range in size from fine sand grains to boulders (reef rock).

As mentioned earlier, rhodoliths are layered coralline crusts that have grown around a center. These centers, or so-called nuclei or cores, may be foreign matter, such as rock, shell, or coral fragments. Alternately, nuclei may be portions of coralline algae. In a Bermuda study Bosellini and Ginsburg[51] found nuclei that were 0.5 to 5 cm thick. Sometimes nuclei are lacking, perhaps because of removal by burrowing animals or because they consisted of soft-bodied animals which decayed. In relict rhodoliths found off the continental shelf off southern South Africa, nuclei were represented by centrally-situated cavities.[404]

FATE OF THE CALCITE

Calcium carbonate accretion in coral reefs is high. In colder parts of the world, where reef-building corals are absent, significant amounts of calcium carbonate may also be deposited. Smith[408] estimated that enormous amounts of calcium carbonate are produced on the mainland shelf of southern California. The calcareous organisms that are responsible are various shell-producing animals and coralline algae in *Bossiella*, *Corallina*, and *Calliarthron*, epilithic nonarticulated forms, and the common epiphyte of surf grass (*Phyllospadix torreyi* Watson), *Melobesia mediocris*. In spite of the preponderance of coralline algae, there are only negligible amounts of calcium carbonate accumulating off southern California. Turnover occurs 0.3—0.5 times per yr for *Bossiella*, *Corallina*, and *Calliarthron* and surprisingly, about 1.5 times per yr for *Lithothamnium* and *Lithophyllum*, and 2 times per yr for *M. mediocris*. On the basis of fouling panel studies, turnover calculations, and estimates from the literature, Smith[408] suggested that the calcium carbonate produced here is flushed into adjacent basins where it partly or completely dissolves.

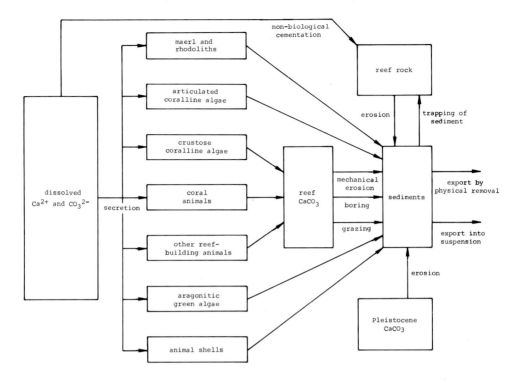

FIGURE 2. A flow diagram of calcium carbonate in a tropical bioherm environment.

Thus far, our discussion has been limited to carbonate deposition in living algae. Of great geological importance in various oceanic environments such as ridges, reefs, and rhodoliths, is the internal carbonate cementation that is not, at least directly, associated with living organisms (Figure 2). The kind of precipitation occurring in spaces or voids in carbonate structures is called cryptocrystallization, and occurs most readily where the local environment is conducive to crystal formation. Primary sources of information on nonliving cementation are Milliman[320] and Alexandersson.[30]

Nonliving carbonate structures of organic origin, such as shells, coral branches and coralline algae, may undergo two transformational processes. The first is mechanical and chemical erosion, and a consequent reduction in size, with the production of dissolved ions and sediments (Figure 2). Relatively large amounts of coral reef sediments may be derived from coralline algae, for example, more than 20% near Bermuda[194] and more than 50% at Johnson Island in the Pacific.[157]

The second process is the growth of calcium carbonate crystals that add to the size or density of a structure. Controlling this nonbiological crystallization are the mineral composition of the original material, the degree of supersaturation of the ambient seawater, and the direct or indirect control effected by organic substances. The greater supersaturation of calcium carbonate in warm seas leads to more crystallization here than in colder waters. In fact, *Lithophyllum* reefs in the Mediterranean are made up of as much as 50 to 70% of cement and internal sediment and relatively little is left of the original algal cell walls.[30]

Cryptocrystallization occurs in coralline algae when carbonate precipitates within empty cells, conceptacles, or in any places where boring organisms have excavated tunnels or galleries. In a living reef, only the surface consists of living tissues forming, as it were, a veneer or blanket partially sealing off the internal nonliving bulk of the reef. The multitudinous chambers and voids are insulated from the surrounding sea and the internal microenvironments are well suited for cementation even where the ambient seawater is undersaturated with calcium carbonate. Hence, submarine cementation tends to occur in living calcareous reefs, and the cementation and lithification are important in producing massive structures.

The predominant cement produced in nonliving coralline skeletons is magnesian calcite. Dolomite [CaMg(CO$_3$)$_2$] and aragonite are also deposited as minor constituents. Alexandersson[30] indicated that aragonite can nucleate directly on crystals of magnesian calcite. Ambient environmental conditions may change with time, and, a common depositional sequence is from magnesian calcite to aragonite. The initial (at least) preponderance of magnesian calcite probably derives from organic substances that originally formed during biomineralization. So, even dead organisms may effect indirect control over crystal formation during inorganic precipitation. Milliman[320] illustrated magnesian carbonate crystals within coralline vegetative cell lumens and also showed recrystallization of magnesian calcite into aragonite.

Strictly speaking, calcite should be more stable in seawater than aragonite.[320] Aragonite is more likely to precipitate inorganically, and it has been suggested that the great abundance of magnesium in seawater plays a significant role in inhibiting calcite nucleation. Aragonite crystallization is not interfered with by magnesium. Favoring the nonbiological production of calcite is the absence of magnesium, low temperature, low pH, and the presence of organic compounds such as sodium citrate and sodium malate. The frequent deposition of aragonite, sometimes even within cavities in dead coralline algae, is favored by the presence of magnesium in solution, high temperatures, and high pH.

The interesting hypothesis that cementation is controlled by extracellular products liberated from living cells at the surface of a reef was presented by Alexandersson.[30] He suggested that extracellular products, which are produced by algae at the surface of the reef, provide a mechanism well suited for influencing the internal chemistry of a reef. In turbulent areas these metabolites may be distributed by water movement into the internal chambers of the reef. Bacterial decomposition of reef organisms may also play a role in providing organic substances that control the type of cement formed. Mineralization, under conditions which would seemingly not be conducive to this process, could still occur. This could explain the observations on marine carbonate precipitation showing that the minerology of the precipitate is independent of the main physiochemical features of the surrounding sea (e.g., salinity, temperature, pH, and ionic ratios). Cementation under indirect control of living cells would explain large biogenic frameworks even in areas where little or no carbonate precipitation occurs in surrounding sediments.

The evidence is circumstantial. Alexandersson[30] pointed out four characteristics of modern carbonate reefs which support the hypothesis:

1. Cementation in living reefs occurs where there is little or no carbonate precipitation in surrounding sediments.
2. Inside reefs "cementation may be varied and ubiquitous, occurring even in the absence of obvious triggering agents".[30]
3. The processes of cementation are extracellular and hence isotopic fractionation would be limited.
4. Local factors contributing to cementation would be living organisms and a pumping mechanism such as found in turbulent areas where pores are present. These features need to be studied.

SUMMARY

When a site becomes occupied by coralline algae, production results in organic and inorganic materials being added to the area.

Growth in coralline algae is slow, with the highest rates in warm waters. Three kinds of measurements may be made, namely, branch elongation rates, marginal extension rates, or increases in thickness.

In many parts of the world, nonarticulated or articulated coralline algae form relatively pure stands that represent climax formations. When surfaces are available and spores are present, new growths may be rapidly initiated.

Organic productivity in coralline algae is lower than it is in many other marine plants and in hermatypic corals. Productivity responds in different ways to light intensity, with the prominent tropical ridge former *Porolithon onkodes* responding positively to high light intensity, in contrast to most other coralline species which are shade adapted.

Inorganic productivity is high for populations of coralline algae, comparing favorably with that for calcifying animals. In tropical reefs calcium carbonate produced by coralline algae have contributed significantly to bioherm structures for millions of years.[451]

Marl and rhodoliths are unattached growths of coralline algae that may accumulate in large amounts. These structures contribute considerably to ocean floor sediments in both cold and warm parts of the world.

The calcium carbonate produced in various calcareous ecosystems may be added to solid structures or may contribute to sediments in nearby areas. In some areas, such as in southern California, the mineral is lost and probably carried away to deep waters. Carbonates are also formed by nonliving cementation in voids in reefs, and this type of calcification contributes substantially to bioherm structures. This secondary crystallization may be under indirect control of living organisms or their excreted metabolites.

Chapter 10

REEF BUILDING

INTRODUCTION

Coralline algae play an important role in building and consolidating tropical reefs — structures projecting toward the water surface in turbulent areas and made up of communities which include lime-producing plants and animals.[391-395] The temporally changing nature of reef systems suggests that their stage of evolution is reflected by their structure and by the kinds of reef-building organisms most prominent in particular localities. Also, reefs may extend up to or above sea level or they may be incipient and well below the surface. Basic similarities occur in both Caribbean and Indo-Pacific reefs, and the differences that have long been recognized may not be directly due to biological or climatic factors but, partly at least, regional peculiarities affected by the rise of the sea. For example, Pacific atolls have developed on slowly sinking volcanoes[320] whereas in much of the Caribbean the reefs grow on Pleistocene platforms.[18]

Many descriptions of tropical reefs are available and recently experimental studies have appeared.[18-19,29,285-287] The long-used classification into barrier (bank-barrier)[18] fringing and atoll reefs still holds, with the last being the most evolved.

From the ecological point of view coralline algae in tropical areas play three roles: (1) primary producers of photosynthate, (2) contributors of sediment, and (3) consolidators and cementers of reefs. The last is the most important, and, in the words of Milliman[320] "This role can be compared to the mortar used in cementing bricks (in this case coral and mollusk debris). In this manner, corallines can consolidate loose grains into hardened rock, and thereby construct stable bioherms."

Tropical reefs occur between 30° north latitude and 30° south latitude[320] although Bermuda is at 32° north latitude and well endowed with reefs due to the northward extension of the warm Gulf Stream waters (Figure 1).

BASIC REEF STRUCTURE

Reef systems consist of structural entities, or *bioherms*[18] as well as associated unstructured communities, such as those dominated by nonreef building coralline algae, corals, angiosperms, or noncalcareous algae. These unstructured communities usually occur behind *algal ridges* extending to the water surface and usually protected from the waves of the open ocean. Bioherms are usually elongated with the long axis at right angles to the incoming waves. Capping many bioherms are algal ridges made up of masses of crustose coralline algae. The distributions of these ridges are related to the strength and constancy of the trade winds.[18-19] The presence of algal ridges in the spectacular Pacific atolls has long been known, but it has only been the recent work by Adey and others that have shown conclusively the presence and importance of coralline algae as ridge-formers in the Western Atlantic. In fact, marked similarities exist in the basic structure of Pacific and Atlantic bioherms, as will be shown below.

A generalized view of a Pacific fringing reef reveals, from sea toward land, a continuous or discontinuous algal ridge parallel to the axis of the waves, a reef flat behind the ridge, a sand-floored lagoon that may or may not be present, and finally, the shore (Figure 2). The ridge usually projects a short distance above low tide level and is exposed, although usually wave-splashed, for short times during some low tides. Doty's[141] "physiological low tide level" is the highest level not exposed to significant drying during low tides. The seaward margin of a ridge slopes into the water until its

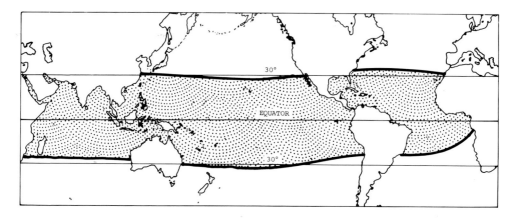

FIGURE 1. Map showing equatorial belt of biotic reefs.

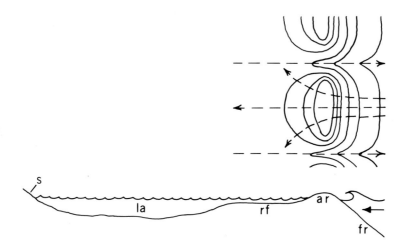

FIGURE 2. Diagram of a section through a bioherm at right angles to the incoming waves showing an algal ridge (ar), forereef (fr), reef flat (rf), lagoon (la), and shore (s). Above is a view looking down on a spur and groove system with depth contour lines. The dashed lines and arrows indicate the direction of water movement inward over a spur and outward through a groove. Adapted from Doty.[141]

surface passes out of the euphotic zone. The landward side of a ridge extends gradually into the shallow reef flat.

Algal ridges are cut by channels at various areas where water that has been thrown over the ridges flows back into the ocean. Doty and Morrison[143] and Doty[141] described some ridges as made up of high incurrent areas (spurs) alternating with low excurrent areas (grooves); the latter are the channels where water moves seaward (Figure 2). The excurrent areas often extend for some distance into the reef flat. The Hawaiian reefs described by Littler and Doty[289] lack the grooves and spurs, the ridges consisting instead of irregularly developed knobs and pillars.

Although the importance of crustose coralline algae as tropical reef builders is now well known, relatively few species are involved. In great tropical oceanic areas only species of *Neogoniolithon*, *Porolithon*, *Lithophyllum*, *Hydrolithon*, and *Sporolithon* are important reef builders and consolidators. These species grow on the reef flats and on wave-pounded algal ridges at just above and below tide level, forming massive pink layers and knobs that cover everything and provide a living, erosion-proof buffer against the waves. In fact, on many viable reefs there is a slow seaward growth of the ridge because accretion is greater then the wear and tear from waves and animals.

GRAZING AND CORALLINE DEVELOPMENT

An important factor helping to determine the dominance of *Porolithon onkodes* and other crustose coralline algae on algal ridges or reef flats is grazing. Filamentous and frondose noncorallinaceous algae that would grow in reef areas are usually visibly lacking except in crevices; they are continuously being removed by fishes and sea urchins. Coralline algae, including *Porolithon,* are also grazed, as revealed by surface scars,[143] but the algal growth exceeds the loss by grazing. Van den Hoek[437-438] suggested that grazing in the *Porolithon*-coral ecosystem on the southwest coast of Curaçao causes it to remain stable and that undergrazing could lead to a gradual deterioration. He further pointed out that the ratio of frondose algae to the *Porolithon*-coral coverage was an indication of the fishing pressure caused by man. Among the most important grazers in the Caribbean are parrot fishes (*Scarus* spp. *Sparisoma* spp.) and surgeonfishes (*Acanthurus* spp.).

Another viewpoint was presented by Adey and Vassar[29] who suggested that grazers restrict the coralline ridge formers to high energy habitats and that were it not for these animals, the algae would play a more prominent role in building tropical bioherms.

In shallow reef communities near Curaçao, extremely heavy grazing may cause destruction of crustose coralline algae.[439] Conclusive evidence for the positive effects of moderate grazing in reef areas was provided by Wanders[442] on the basis of caging experiments at Curaçao. His findings may be summarized as follows:

1. Living plants of *Porolithon pachydermum* and *Lithophyllum* sp. placed in cages soon become overgrown by mats of delicate algae such as *Giffordia, Enteromorpha,* and *Cladophora* and eventually other algae, including the coralline *Jania capillacea.*
2. The overgrown coralline algae died.
3. Boring algae, such as *Ostreobium quekettii* Born. et Flah. and *Plactonema terebrans* (Born. et Flah.) Gom., were present in the dead crusts.
4. Extremely intensive grazing may damage reef building coralline algae.
5. The main grazers are fish and the sea urchin *Diadema.*
6. Grazing is necessary for the settlement and early growth of coralline spores.

It seems that grazing promotes the establishment and growth of coralline algae.

In addition to the removal of calcium carbonate by grazing fish, there are several organisms that bore mechanically or chemically into a reef. In fact, the carbonate may be riddled with cavities of boring macroorganisms such as clinoid sponges, polychaete worms, bivalves, and microorganisms, such as endolithic algae and fungi.[30] This weakens the structure and provides the chambers and cavities where cryptocrystallization occurs. The roles that these organisms play are largely unknown, but certainly reefs that are honeycombed by these borings degenerate rapidly. A boring echinoid, *Echinometra,* is conspicuous in Caribbean reefs[29] and bluegreen algae, such as *Ostreobium* in Bermuda, form micro tunnels inside the carbonates. Littler[288] mentioned that his unidentified melobesioid C in Hawaii might be an example of a rockboring rhodophyte.

Table 1
THE PERCENT COVERAGE OF THE MAIN REEF-BUILDING
CORALLINE ALGAE ON THE REEF STUDIED BY LITTLER AND
DOTY AT WAIKIKI IN THE HAWAIIAN ISLANDS[a]

Organisms	Percent cover	Comments
Hydrolithon reinboldii	10.6	As crusts or nodules throughout reef except at seaward edge of ridge
Sporolithon erythraeum	5.7	In shade throughout reef flat; coincides with dense *Sargassum*
Porolithon onkodes	3.0	Almost all in seaward edge of algal ridge
P. gardineri	0.8	Almost all in grooves passing through ridge
Lithophyllum kotschyanum	1.0	Included with *P. gardineri* before Littler and Doty[289]
Other crustose coralline algae	0.8	
Total of coralline algae	21.9	
Coral animals	<0.2	Widely scattered small colonies
Dead reef, sand, and rubble	~6	

[a] Adapted from Littler[285] and Littler and Doty.[289]

INDO-PACIFIC REEFS

Setchell[391-393] argued that coralline algae are important reef builders and consolidators in the Indian and Pacific Oceans. This is now known to be true. For example, *Neogoniolithon frutescens, Lithophyllum kotschyanum,* and three species of *Porolithon, P. onkodes, P. gardineri,* and *P. craspedium,* are ridge formers and, in Hawaii *Sporolithon erythraeum* and *Hydrolithon reinboldii* are important in consolidating reef materials.[285]

In their major aspects, algal ridge communities are similar throughout the Indo-Pacific. Littler[285] listed publications dealing with ridges at Bikini Atoll, Indian Ocean Atolls, Solomon Islands, Mopelia Atoll of the Society Islands, Tutuila and Tahiti, Funafuti Atoll, Raroia Atoll, and Ronglap Atoll. In all these reports prominent ridge roles are ascribed to *Porolithon* and the unity of the structures is apparent. Publications by Littler and Doty have revealed much about the role that crustose coralline algae play on a fringing reef in Hawaii.

Extending offshore from the War Memorial Natatorium at Waikiki Beach, Oahu Island, is a fringing reef complex dominated by nonarticulated oralline algae.[285] Five species of encrusting coralline algae are important reef builders and consolidators and nine others are present (Table 1). There are two species of *Porolithon;* they require specialized habitats and are discussed later. *Hydrolithon reinboldii* and *Sporolithon erythraeum* occur over much of the reef flat as well as on the forereef until they disappear in deeper water. *Hydrolithon reinboldii* and *S. erythraeum* grow well in shade and cement large, loose rocks to the reef surface. Coralline algae cover 21.9% of the

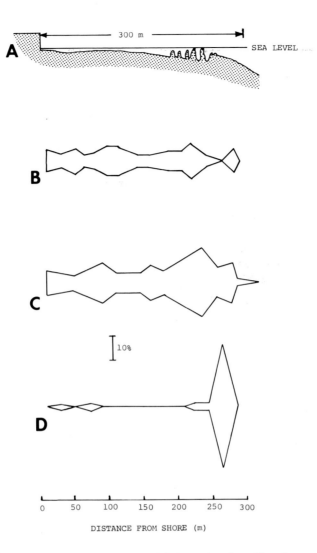

FIGURE 3. Percent coverage of three species of coralline algae at the Waikiki reef studied by Littler.[285] A. Profile of the reef from the War Memorial Natatorium to the forereef. The algal ridge here consists of knees. B. *Sporolithon erythraeum*. C. *Hydrolithon reinboldii*. D. *Porolithon onkodes*. Note that *P. onkodes* predominates on the ridge. Adapted from Littler.[285]

reef surface; the cover provided by three species is shown in Figure 3. It is interesting that the coelenterate coral cover on the Waikiki reef went from 4.6% in 1928[359] to less than 0.2% at the time of Littler's study in the late 1960s. Littler[280] suggested that sewage-derived phosphates may be the reason for the increased ratio of coralline to coral cover. Some coral animals are extremely sensitive to increased phosphates whereas some coralline algae apparently grow well in these types of environments. Specimens of *H. reinboldii* sometimes appeared to be growing well even though covered by a layer of sandy silt.

POROLITHON IN THE PACIFIC

The most important Pacific and Indian Ocean reef-building coralline is *Porolithon onkodes*, the so-called "pavement nullipore."[394] Neither it nor the other two species treated below have been reported from the Atlantic Ocean, although *P. pachydermum* is an important reef builder in parts of the Caribbean Sea.[437-438] In the Pacific, *P. onkodes* often occurs in the same reefs as two other species in the genus, *P. gardineri* and *P. craspedium*, for example, at Rongelap Atoll in the Marshall Islands.[266]

Porolithon onkodes and *P. gardineri* are prominent on the fringing reef complex at Waikiki Beach and cover 3.0 and 0.8% of the total reef area, respectively.[289] These species, in association with *Lithophyllum kotschyanum* (1.0% cover), especially dominate the seaward margin or ridge of the reef where it forms irregular columns or knees (Figure 3). Usually *P. onkodes* is more widely distributed over the ridge than is *P. gardineri*.[289] However, subtidally in the crest region, the highly branched *P. gardineri*, with a cover of 6%, and *L. kotschyanum* are prominent. *Porolithon onkodes* largely replaces these species in the intertidal zone on the crest of the ridge and down the seaward slope with the upper parts of the knees completely covered by this species. In the areas of greatest wave shock, *P. onkodes* forms superimposed crusts up to several centimeters thick, but it becomes thinner until it is only a few millimeters thick where it disappears at a depth of 5 to 10 m on the forereef. The relative positions of the two species of *Porolithon* may also be seen in the fact that *P. gardineri* reaches its maximum frequency in areas of strong wave-action 250 m from shore whereas *P. onkodes* reaches its maximum 20 m farther seaward. The description of this reef by Littler and Doty[289] revealed that *P. onkodes* has a higher cover, density, and frequency than *P. gardineri* and this, coupled with its position on the reef front, makes it the ecologically most important coralline species in building and maintaining the reef edge. This appears to also hold in many other Pacific and Indian Ocean reefs.

An interplay of factors, such as wave force, grazing, and shading, appears to result in the continued dominance of *P. onkodes* on the algal ridge and down the reef front. Littler and Doty[289] reported on clearing experiments and growth on glass panels installed on the ridge as well as on the shallow reef-flat. Other plants that are early, successional stages in experimental areas are species of *Calothrix, Sargassum, Jania, Peyssonnelia, Sphacelaria,* and *Padina*. Under conditions of increased shading and decreased grazing, these plants prevented the successful establishment of *P. onkodes* in denuded areas and encroached upon and killed the coralline in adjacent areas. Wave shock on the ridge crest and grazing by fish and sea urchins on the outer slope repress the frondose algae and are partly responsible for the climax reef community dominated by *P. onkodes*.

CARIBBEAN REEFS

The question of whether or not algal ridges occur in the Caribbean has had a mixed history, with statements to the affirmative made as long ago as 1906[165] (with reference to reefs of *Lithophyllum* in Puerto Rico), and to the negative as recently as 1969.[417] Now, papers on reefs in the Caribbean Sea are appearing at an ever increasing rate, and many of them contain substantial amounts of information on coralline algae.[16,18-19,22,416,437-438,440] Van den Hoek and his co-workers described the zones of marine vegetation in Curaçao in the Netherlands Antilles, and Adey and his co-workers have focused on descriptive and experimental studies of reefs at St. Croix in the U.S. Virgin Islands.

Adey[19] mapped the distribution of algal ridges in the West Indies by aerial reconnaissance, finding that up to an elevation of 300 m the ridges could easily be seen (Table 2). After correlating their distribution with prevailing winds, he found that their positioning is directly related to the winds and wave movement. In the Caribbean Sea the most successful ridges are those exposed to the prevailing easterly trade winds.

Table 2
ALGAL RIDGE AREAS OF THE CARIBBEAN SEA AND WEST INDIES. DATA FROM ADEY[19]

Region	Location	Development and extent
Caicos Hispaniola	Plana Cays	Linear series of cup reefs and at least one linear ridge; medium height
	Great Inagua	Abundant, scattered cup reefs and at least one short linear ridge, in a restricted area; low height
	N. Hispaniola	Short and linear, especially along channel edges in large linear reef systems; low height
Antilles	E. Hispaniola	Several sublinear features in reef complex at mouth of Samana Bay; scattered cup reefs along northeastern shore; low height
	N.E. Puerto Rico	Numerous cup reefs and a few sublinear ridges east of San Juan; low height
	Anegeda	Abundant sublinear features especially along channels in major reef complex; medium height
	Anguilla/St. Martin	Scattered cups in Prickly Pear Cays, north Anguilla and southeastern St. Martin; low height
Lesser Antilles	St. Croix	Scattered cup reefs to sublinear complexes; medium height to degenerate
	St. Eustatius	Two short sublinear features near easternmost point, high
	Barbuda	Very abundant cup reefs in northeast and abundant sublinear ridges in main eastern reef; high
	Antigua	Scattered cup reefs from Dian Point to Willoughby Bay on east coast; low
	Grande Terre	Major linear algal ridge system extends along much of the northern part of the eastern coast, very high
	Desirade	Abundant sublinear ridges, cup reefs and algal lips along entire shore; medium
	Marie Galante	Major sublinear and cup reef systems along southeastern shore; very high
	Martinique	Abundant sublinear ridges in southeast; more scattered in east; high

Table 2 (continued)
ALGAL RIDGE AREAS OF THE CARIBBEAN SEA AND WEST INDIES. DATA FROM ADEY[19]

Region	Location	Development and extent
South	Grenadines	Scattered small cups in eastern Canouan and southern Carriacou; low
	Los Roques	Very few short sublinear features in northeastern area; low
	Puerto Cabello	Short, well-developed, sublinear ridge; medium to high
South West	San Blas	Extensive high sublinear ridges in Holandes Cays; becoming lower and more scattered to both east and west
	San Andres	One short sublinear ridge; medium height
	S.E. Jamaica	One short section of cup reefs, medium height
North West	Cozumel	Series of cup reefs in northeast; low height

Most Caribbean algal ridges originate by the fusion of mushroom-shaped cup reefs which represent earlier stages of development. Algal cup reefs (also called microatolls and boilers) also occur in Bermuda where they are a notable feature of the south shore.

Meischner and Meischner[315] have outlined the Holocene development of the cup reefs in Bermuda as they formed during the rapid rise in the sea level during the last 10,000 years. Here, as in the Caribbean, there are three Pleistocene terraces presently located at depths of 35 m, 20 m (the most extensive terrace), and probably 8 m (Figure 4). As the sea rose the first algal cup reefs developed on the edges of the 20 m terrace creating a lagoon that eventually became a sandy moat. These reefs died and do not now reach the water surface. A second line of algal cup reefs developed on the 8 m terrace. These cup reefs reach the present sea surface and are still living. A third series of cups has now been established near the shore, and Meischner and Meischner[315] suggested that these reefs will eventually function as a fringing reef system behind which a new lagoon will develop.

Cup reefs also occur in the Caribbean, for example at St. Croix.[22] Here, rapid marginal growth of bioherms often leads to their formation and, in time, they tend to fuse irregularly to form crude algal ridges (Figure 5). The relative strength of the wave onslaught determines the rates with which these cups fuse and form discrete ridges as they move seaward. In the most wave-intense areas growth seaward is relatively rapid and old cups may be left behind an advancing ridge.

In a recent overview of reef structures, Adey[18] focused on the dynamics of reef morphogenesis in the Holocene as related to wave energy and sea level rise. It seems that these two factors, plus the presence of limestone shelves formed millions of years ago, must be considered in order to understand reef development in the Caribbean Sea.

Adey and Burke[22] pointed out that the volcanic activity that had marked the Lesser Antillean island chain in the last 30 million years has negated the establishment of extensive carbonate shelves around some islands, hence major reefs are absent (i.e.,

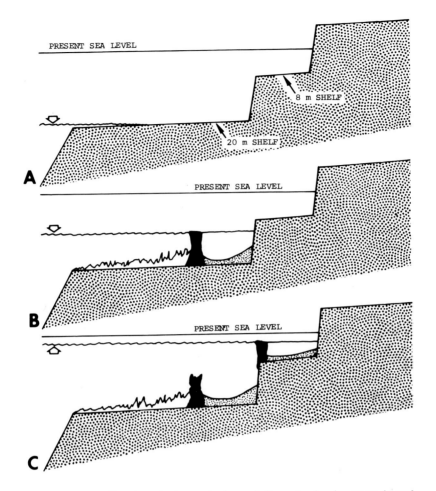

FIGURE 4. Profiles through the south shore of Bermuda showing stages in reef growth (black). Arrows show sea level. A. About 9000 years BP. B. About 7000 years BP. C. About 3000 years BP. Adapted from Meischner and Meischner.[315]

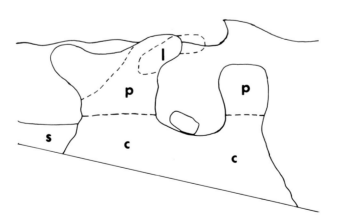

FIGURE 5. Profile through an algal ridge formed from fused cup reefs. Note that an intertidal rim is growing seaward and that a previously formed rim has broken off. l = *Lithophyllum*; p = *Porolithon*; c = coral animals; s = sand. Adapted from Adey.[19]

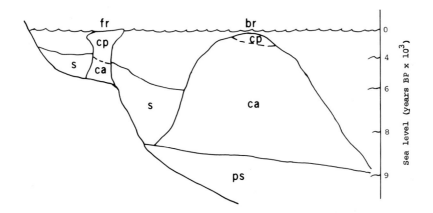

FIGURE 6. Generalized diagram of a section through a fringing reef (fr) and bank-barrier reef (br) in a Caribbean bioherm such as at St. Croix. cp = coralline plants; ca = coral animals; ps = Pleistocene shelf; s = sand and sediment. Adapted from Adey.[18]

at St. Vincent, northern Martinique, Dominica, Montserrate, and Saba). Around others, particularly those that are small and low with little rainfall and sediment runoff, there tends to be good reef development (e.g., Anegada, Barbuda, Antigua, St. Croix, Grande-Terre, and the southern Grenadines). Probably of critical importance in determining the extents of reef systems in the eastern Caribbean are the heights of the Pleistocene shelves. Windward shelves with present depths of less than 15 to 18 m developed massive coral bioherms and sometimes algal ridge systems. These structures are absent in deeper water. Viable algal ridges have also been described from northeastern Brazil, Venezuela, eastern Yucatan, Bermuda, and other places in the western Atlantic.

The St. Croix reef system consists of three parallel bioherms.[18] Close to shore are intermittent fringing reefs, farther out are bank-barrier reefs, and several kilometers from the shore are submerged reefs. The first two are capped by coralline ridges (Figure 6). The fringing reefs began to grow 5000 to 3000 years BP as corals (the antler coral *Acropora palmata*) formed incipient bioherms which became capped by crustose coralline algae. The subsequent development of the protecting bank-barrier reefs, coupled with a declining rate of sea-level rise, resulted in the algal ridges being in various stages of degeneration. The extent of algal ridge development seems to be directly related to the extent of the wave energy to which it is exposed. Coralline algae are favored in high-energy areas, but they build bioherms more slowly (3 to 6 m vertical growth per 1000 years) than do coral animals (8 to 15 m/1000 yr). Adey[18] pointed out, that the total vertical growth of Holocene reefs in the Caribbean is probably more than two times as great as for Pacific atolls.

CORALLINE RIDGE-FORMERS IN THE CARIBBEAN

As *Acropora* growth brings mounds or pillars of a reef to within 2 to 6 m of the water surface, crustose coralline algae in *Porolithon* and *Neogoniolithon* replace the animal structures.[22] As development continues and these structures approach mean low water, the important Caribbean framework builder, *Lithophyllum congestum* (also called *L. daedaleum*), often becomes a major element, especially along the margins of the mounds or pillars.

The ecological relationships occurring in the various habitats in the St. Croix reef system have been studied by Adey and Vassar.[29] They planted clean surfaces of rough polyvinyl chloride (PVC), some of which were constructed so as to resemble branches of the coral *Acropora palmata*. Crustose coralline algae grew well on the surfaces, although the presence of noncorallinaceous crustose algae, limpets, parrot fish, and other organisms affected the development of the corallines. Early in ridge development *Neogoniolithon megacarpum* and *Porolithon pachydermum* dominate if occasional calm water allows for the grazing of fleshy algae. In partial shade occur *Neogoniolithon accretum* as well as noncorallinaceous algae and animals, including corals. When a developing algal ridge reaches sea level *Lithophyllum congestum* replaces the other organisms and builds rapidly to 20 to 30 cm above sea level. Above 30 cm *P. pachydermum* tends to replace *L. congestum*. These two algae are responsible for the bulk of the algal ridge. The blocking of waves leads to the eventual degeneration of algal ridges, a development sometimes hastened by boring animals.

Lithophyllum congestum dominates ridge-edge habitats off northwest Puerto Rico, southeastern St. Thomas, northeastern Virgin Gorda, southeastern St. Martin, eastern and southern St. Eustatius, eastern Antigua, Guadeloupe, southeastern Martinique, and eastern St. Croix.[416] Apparently its northern boundary is Cuba. Interestingly, Steneck and Adey[416] surveyed several other areas from southeastern Florida through the Bahamas without finding *L. congestum*.

L. congestum requires strong to medium wave action and has a narrow vertical range, with most coverage being only +0.1 to −0.1 m (with reference to mean low water) although it does occur as high as +0.2 m and as low as −3.0 m. Hence, it covers very little reef area in spite of its prominence in the Caribbean. In fact, this narrow vertical distribution has been useful in determining ancient sea levels since this entity is often present in cores as deep as 4 m below the surfaces of algal ridges.[16] It is this restricted coverage of reef areas that perhaps relegates it second to *Porolithon pachydermum* in total coralline carbonate accretion in the area.[16] On ridges receiving extreme wave action in the easternmost Lesser Antilles, this species is replaced by *Neogoniolithon* as the major builder.[22] Moreover, because this species tended to be replaced by *P. pachydermum* in high ridges off St. Croix, Steneck and Adey[416] suggested that the latter entity is more desiccation-resistant. *Lithophyllum congestum* is structurally and ecologically related to *L. kotschyanum* in the Pacific.

On the shores of Curaçao the most prominent reef-building coralline species is *Porolithon pachydermum,* which forms thick crusts at slightly above to just below mean low water in areas of high wave energy. Roughly speaking, there is a decrease in the amount of and thickness of this smooth crustose species from the most wave-exposed to quieter shores where it finally disappears in bays and inlets.[437] *P. pachydermum* may, in some respects at least, be considered the Caribbean equivalent of *P. onkodes*, and it appears to play an important role in reefs and along shores where wave energy is very high.[19,437] It is the most prominent crustose coralline at about low tide level in Curaçao,[437] but, on the reefs off St. Croix, it plays a secondary role to *Lithophyllum congestum*. Possibly the latter species outcompetes *Porolithon pachydermum* in areas of slightly lower wave energy and lower light intensities,[416] although other factors may be involved.

As may be seen in Figure 7, a common coastal configuration on wave-eroded coasts of Curaçao consists of steep cliffs with wave-caused "surf-platforms" with upper surf-niches that are undercut by lower "surf-niches."[437] Van den Hoek[437] tentatively concluded that the upper surf-niches are caused by dissolution of the limestone by splash and the lower surf-niches by mechanical abrasion by boulders and sand (Figure 7). Mean sea level (MSL) (the tidal range here is about 30 cm) is just below the seaward edge of the surf-platforms, and it is about 1 m below MSL where *P. pachydermum* forms its most massive growths. This so-called *"Porolithon-cap"* may be primarily responsible for the formation of the surf-platforms by growing in the zone of maximum turbulence and serving to prevent erosion of the underlying limestone at that

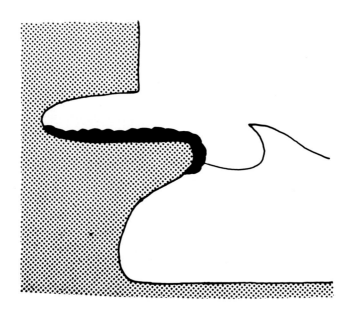

FIGURE 7. Profile of a wave-exposed cliff in Curaçao, Netherlands Antilles. A surf platform is covered with *Porolithon pachydermum* (black). The upper surf niche is probably caused by dissolution and the lower niche by mechanical abrasion. Adapted from van den Hoek.[417]

level. The lower surf-niches are extensions of large submarine plateaus which, at a depth of about 5 m, reach at least 100 m from shore. *Porolithon pachydermum* in Curaçao has a rather wide ecological amplitude with regard to wave exposure and depth.

Taylor[432] recorded two other species of *Porolithon* from the Caribbean, but they are apparently rare and in need of verification. Dawson[109a] described one from the Gulf of California. Thus, there are five species of *Porolithon* that are prominent in certain reefs, three from the Pacific and Indian Oceans, one from the Gulf of California, and one from the Caribbean.

SUMMARY

Bioherms are massive structures built by calcareous organisms in tropical areas of the world. For a long time they have been recognized as barrier, fringing, and atoll reefs. Coralline algae are important in producing organic matter, contributing to sediments and, especially, in cementing and consolidating the bioherms.

From the open ocean shorewards, a fringing reef system typically consists of fore-reef, algal ridge, reef flat, lagoon, and shore. Reef building coralline algae comprise most of the mass in algal ridges where they outcompete other organisms in building the structure in the face of onrushing waves. Consolidating crustose coralline algae and articulated turfs are present behind the ridges, particularly in the reef flats. Algal ridges extend up to and above sea level where the species that are adapted to turbulent, well-lighted conditions are few, belonging mostly to *Porolithon* and *Lithophyllum*.

Fishes, sea urchins, and limpets eat the small fleshy algae that inhabit bioherms, especially behind the algal ridges. Grazing is important in preventing the coralline algae from becoming overgrown by these algae. Grazing is low in wave exposed ridges, and here coralline development is most pronounced.

In Indian and Pacific Ocean atolls and in fringing reefs in Hawaii species of *Porolithon*, *Lithophyllum*, and *Neogoniolithon* are the main ridge-formers. In Hawaii species of *Hydrolithon* and *Sporolithon* grow behind the ridges and consolidate reef materials. In the fringing reef off Waikiki Beach crustose coralline algae account for nearly 40% of the total cover, a figure that may be higher now than previously because of pollution and a decrease in the numbers of coral animals.[285]

The pavement nullipore, *Porolithon onkodes*, and the knobby *P. gardineri* and *P. craspedium* build massive ridges or pillars in wave exposed, well-lighted areas. The smooth *P. onkodes* produces superimposed crusts up to several centimeters thick where the wave onslaught is greatest.

Numerous algal ridges produced by *Lithophyllum congestum* and *Porolithon pachydermum* have recently been found to occur in the western Atlantic. Caribbean bioherms began as coral animal structures building on Pleistocene shelves as sea level rose in the Holocene. As these structures approached the sea surface, incipient algal ridges appeared, usually as small algal cup reefs. Under the right conditions of wave action, these coalesced to form the ridges. In reef systems in the Caribbean there are often three parallel structures, namely, fringing reefs close to shore, bank-barrier reefs farther out, and submerged reefs several kilometers from shore. The algal ridges occur on the first two.

Lithophyllum congestum grows in a narrow vertical range centered at mean low water, and *Porolithon pachydermum* can grow higher in the intertidal, being apparently desiccation-resistant. In Curaçao, *P. pachydermum* forms horizontal edgings on cliffs above which surf-platforms are cut and below which surf-niches are eroded.

Chapter 11

FOSSIL CORALLINE ALGAE

INTRODUCTION

As would be expected, the hardness of coralline thalli has resulted in their ready fossilization and there is a voluminous amount of literature on fossil coralline algae. Much of this literature is in geological journals not usually seen by biologists. Hence, there are two schools of research on coralline algae, that on fossil and that on extant plants. Unfortunately, there has been relatively little cross-communication between these two areas, and researchers in one area tend to know little about the other. Much of the work on fossils has been concerned with describing plants that grew in different geographic areas during different geological ages. Some describing has been done with material that shows imperfectly the structures needed to assign plants to genera and species. This, coupled with a communication gap between those working with classification systems involving fossil plants and those working with extant plants, places doubt in many of the generic concepts recognized by the two groups of scientists. The placement of fossil taxa in extant genera, listed in Tables 3 and 4, Chapter 1, must be examined with caution. Also, care must be taken in relating fossil taxa to classification systems erected on the basis of living plants, such as the system given in this book.

Part of the problem, of course, is the fact that paleobotanists must work with plant tissues or plant impressions that are millions of years old. Cellular features that may be essential, such as secondary pit-connections and epithallia, are not preserved. Classification must be based on conceptacle type and location, hypothallial (and medullary), and perithallial (and cortical) characters, and megacells. As a result Wray[452] and others classify fossil Corallinaceae into the old subfamilies Melobesioideae for crustose taxa and Corallinoideae for articulated forms.

The imperfect nature of fossils has caused difficulties in assigning material in genus. For example, Wray,[451-452] related how *Keega*, a late Devonian genus from Australia, thought to belong to the Solenoporaceae, was reinterpreted in 1974[483] as the basal layers of fossil stromatoporoids. Wray,[452] the author of the genus, concurred with this reinterpretation. Primary distinctions among extant *Lithothamnium*, *Phymatolithon*, and *Clathromorphum* are based on uncalcified structures, that is, on epithallia and reproductive cells.[13] *Lithothamnium* may include fossil taxa which would have been assigned to one of the other genera if they could have been examined critically. *Lithothamnium*, as well as *Lithophyllum*, have long been repositories for a wide assortment of fossil crustose coralline algae.

Obviously, the difficulties in understanding fossil species are even greater, and Wray's statement[452] sums it up: "... the literature on fossil corallines has been cluttered with many useless species based on superficial characteristics and minor variations in dimensional data."

Four families of calcareous red algae containing fossils are presently generally accepted.[452] They are as follows: Solenoporaceae, Gymnocodiaceae, Squamariaceae, and Corallinaceae. The first two are extinct, but the last two have extant representatives. We shall exclude the noncorallinaceous Gymnocodiaceae, a family containing entities resembling Recent *Galaxaura* (Chaetangiaceae) and the Squamariaceae, which may, however, have been derived from solenoporacean entities in the Paleozoic.[451-452]

SOLENOPORACEAE

Erected by Pia,[484] this extinct family contains 10 to 12 genera of crustose, leaf-like

Table 1
SOME GENERA IN THE SOLENOPORACEAE

Genera	Times
Solenopora Dybowski 1878	Late Cambrian, Ordovician, Devonian
Pseudochaetetes Høug 1883	Ordovician
Pycnoporidium Yabe et Toyama 1928	Permian, Jurassic
Parachaetetes Deninger 1906	Ordovician, Devonian
Neosolenopora Mastrorilli 1955	Miocene (?)
Solenoporella Rothpletz 1891	Ordovician

or, more often, nodular plants made up of filaments of broad cells (Table 1). The longitudinal walls between contiguous filaments are conspicuous, but the preservation of transverse walls varies, and these "partitions"[451-452] are absent or nearly so in *Solenopora*, but well-defined in *Parachaetetes*. The characteristics that set the Solenoporaceae apart from the Corallinaceae are (1) vegetative cells 30 to 60 μm broad, (2) crusts not, or only weakly, distinguishable into hypothallia and perithallia, and (3) the rare occurrence of reproductive structures.

The Solenoporaceae appeared first in the Cambrian Period and, by the Ordovician Period, these plants were abundant, following which their numbers apparently diminished until they again became important in the Jurassic.[450] *Solenopora* and *Parachaetetes* were common and widespread in the last half of the Paleozoic Era, with the former first recorded from the Cambrian and both genera existing at least until the Paleocene, when the family became extinct (Figure 1).

The primary characteristic setting this family apart from the Corallinaceae is the large cell diameter. Also, the reproductive structures are rarely present, and it is presumed that most species produced external and uncalcified spores. An exception is *Neosolenopora armoricana* for which Elliott[153] described fossils which are vegetatively solenoporoid, but in which there were strata of oval cavities 182 μm high × 109 μm in diameter that were probably *Sporolithon* type conceptacles. Since these fossils are from the Miocene at a time when Corallinaceae were numerous, Elliott suggested that by this time some solenoporacean species produced spores in calcified chambers which the early taxa lacked. He further suspected[488] that the Solenoporaceae did not evolve past the Paleocene because of their large cell size that may have reflected a basic structural and physiological schism between them and the more successful Corallinaceae. Possibly *Neosolenopora* is among the last solenoporacean corallines and the only one having carbonate-encased reproductive structures.

An example of the difficulties that may be encountered when working with fossil specimens is that experienced by Elliott[153] when he described *Cretacicrusta dubiosia* from the Cretaceous. This crustose species had (1) a clearly differentiated hypothallus and perithallus and (2) a perithallus consisting of broad "septate-tubular" filaments 100 to 180 μm in diameter. Thus it had a corallinaceous characteristic (No. 1) as well as a marked solenoporacean characteristic (No. 2) and could not even be placed in one or the other family. Wray[451-452] suggested that the Solenoporaceae was comparable to the Corallinaceae with respect to environment except that the former did not live in as wide a range of depth and temperature. Paleobotanists are in general agreement that cellular organization in the Solenoporaceae is primitive and mostly distinct from that in the Corallinaceae, although there are exceptions to this (e.g., *Cretacicrusta*). Prob-

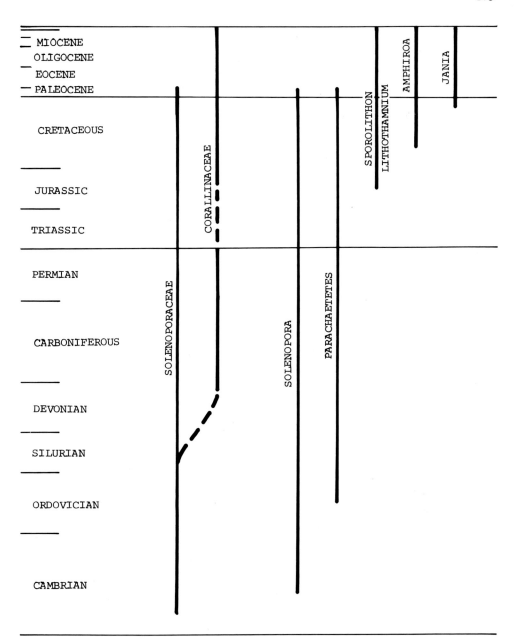

FIGURE 1. Geological ages of the Solenoporaceae and Corallinaceae and some common genera. Adapted from Wray.[452]

ably elements of the Solenoporaceae are ancestral to the Corallinaceae, although direct evidence for this is absent.

ANCESTRAL CORALLINE ALGAE

In addition to the Solenoporaceae, another group of calcareous red algae was common in the late Paleozoic. Wray[450] called them "ancestral corallines." This artificial group contains plants that are structurally similar to recent Corallinaceae, but they are grouped separately because of a lack of representation during the Triassic before un-

Table 2
SOME GENERA OF PROBABLE CORALLINACEAE KNOWN ONLY AS FOSSILS

Genera	Time
Aethesolithon Johnson	Miocene
Archaeolithoporella Endo	Cretaceous
Archaeolithophyllum Johnson	Late Paleozoic
Cretacicrusta Elliott	Lower Cretaceous
Cuneiphycus Johnson	Late Paleozoic
Dasyporella Stolley	Ordovician
Hemiphyllum Lemoine	Cretaceous
Hensonella Elliott	Cretaceous
Katavella Chuvashov	Devonian
Kymalithon Lemoine et Emberger	Cretaceous
Palaeothamnium Conti	Miocene
Paraphyllum Lemoine	Cretaceous
Paraporolithon Johnson	Oligocene
Petrophyton Yabe	Cretaceous
Pseudoaethesolithon Elliott	Miocene
Pseudolithothamnium Pfender	Cretaceous
Stenophycus Fenton	Devonian

doubted Corallinaceae first appeared in the Jurassic (Table 2). A genus in this group, *Archaeolithophyllum*, was an important component of Late Paleozoic marine floras beginning in the Devonian.[450,486] Among the ancestral corallines, this genus is the most widespread both geographically and stratigraphically. *Archaeolithophyllum* closely resembles *Lithophyllum* except for large-celled hypothallia. The "ancestral corallines" were important reef builders in Upper Carboniferous and Lower Permian times[450] with phylloid or leaf-like forms of *Archaeolithophyllum* and *Cuneiphycus* binding and cementing the reefs in much the way crustose coralline algae do today.

APPEARANCE OF THE CORALLINACEAE

The Corallinaceae first appeared in the Jurassic and evolved rapidly in the Cretaceous and early Cenozoic and several recent genera have been reported from these times (Table 3). Wray[450] believed that the ancestors of this family are among the ancestral coralline algae which, in turn, may have evolved from the Solenoporaceae (Figure 1). Several evolutionary patterns based largely on fossils have been suggested.[451-452] For example, one idea is that of deriving the melobesioid algae from *Solenopora* in the early Mesozoic, and the lithophylloid and articulated forms from *Parachaetetes* by

Table 3
EXTANT CORALLINE GENERA ALSO REPORTED AS FOSSILS. MOSTLY FROM WRAY[452]

Genera	Earliest fossil
Lithoporella	Late Jurassic
Lithothamnium	Late Jurassic
Sporolithon	Late Jurassic
Lithophyllum	Early Cretaceous
Amphiroa	Cretaceous
Arthrocardia	Cretaceous
Jania	Cretaceous
Mesophyllum	Paleocene
Corallina	Eocene
Tenarea	Eocene
Melobesia	Oligocene
Calliarthron	Miocene
Neogoniolithon	Miocene
Porolithon	Miocene

way of the ancestral corallines in the Late Paleozoic.[450] However, schemes such as this need to be reworked with greater consideration for phylogenetic clues provided by extant coralline algae.

A trend that is more certain to have occurred during past geological time is evolutionary success for those coralline algae having small vegetative cells; now most coralline algae have cells less than 15 μm in diameter. It is tempting to speculate that, since the calcium carbonate is present in cell walls, having smaller cells leads to more lime per unit volume and greater success because of a harder thallus. This is certainly an oversimplification, but nevertheless, the trend is there.

The most advanced types of coralline algae — the articulated forms — are also present in the fossilized state (Table 2), being particularly common in the Tertiary. As with other fossil forms, placement into genus is often questionable.

A unique, jointed coralline genus, *Subterraniphyllum*, was described from the Oligocene, Eocene, and Miocene by Elliott.[150] The large-celled, tiered medulla was composed of cells 78 μm in diameter and 117 μm long, whereas the cortical cells were only 10 to 15 μm in any dimension.

SUMMARY

Studies of the structure and taxonomy of coralline algae have evolved along two lines: (1) studies of fossil coralline algae by paleobotanists and (2) studies of living plants that were fixed while fresh. Unfortunately, the two groups of researchers have interacted very little. Hence, there are large discrepancies in the taxonomic limits of several genera and subfamilies. The literature on fossil coralline algae is large, and mostly in geological journals.

The Solenporaceae is a family containing several extinct genera, some of which are probably ancestral to the Corallinaceae. In *Solenopora* and *Parachaetetes,* two genera in the Solenoporaceae, the vegetative cells have a larger diameter than in the Corallinaceae; they are more than 30 μm broad. In most solenoporacean fossils reproductive structures are lacking, probably because they were borne on the thallus surface and not preserved. This family is well represented in the Paleozoic and even has been found as late as the Miocene.

Fossil coralline algae structurally similar to extant Corallinaceae are present in the Late Paleozoic. However, they are absent in the Triassic and therefore have been grouped separately from this family, which first appeared in the Jurassic and continues to the present.

Numerous fossil species have been placed in genera of living coralline algae, but some of the placement is in doubt because of the fragmentary nature of the material. Unusual fossils have been described, for example, *Cretacicrusta,* which combines characteristics of the Solenoporaceae and Corallinaceae, and *Subterraniophyllum,* an articulated entity with enormous medullary cells.

Chapter 12

TAXONOMY

INTRODUCTION

Because so many researchers have contributed to the taxonomy of coralline algae it is difficult to organize a coherent unbiased discussion. Most authors that present information on structure also give their views on how their findings should be used in taxonomy. As in most other red algae, coralline taxonomy is based on descriptions of vegetative and reproductive structures. The data on coralline structures are vast and include information on form, tissue anatomy, reproductive organs, sporeling development, and cellular structure.

The emphasis in this chapter is on four areas of taxonomy:

1. The history of the study, naming, and grouping of the taxa.
2. The characteristics that are used to delimit taxa higher than species.
3. The identification of taxa of coralline algae.
4. Suggestions as to the direction that evolution has taken in arriving at the coralline algae present in the oceans today.

As each of these topics is discussed the focus will be on genera and higher taxonomic catagories.

Most descriptions of species of articulated coralline algae rely heavily on frond shape, particularly branching characteristics, and intergenicular length, width, and shape. Even into the 20th century, small differences in these features were used to delimit species. Because of a paucity of information on the extent of variation in size and shape, far too many species have been described. Today there are more than 400 species and varieties of articulated coralline algae. The problems in deciphering the taxonomy and nomenclature of coralline algae are many.

HISTORY

Several papers provide insight into the development of our early knowledge of coralline algae: Suneson,[421] Manza,[301] Johansen,[223,231] Cabioch,[76] Littler,[284] Adey and MacIntyre,[24] Gittins,[187] and Lebednik.[263]

Most authors writing in the 1700s about what we now treat as coralline algae considered them to be animals, or at least animal-like (Table 1). Apparently the calcareous nature of the organisms and the presence of small conceptacle pores induced these scientists to relate them to coral animals. Ellis[155] had a strong influence on contemporary biology, and he arranged them with corals. Linnaeus[280-281] followed Ellis' opinion on this when he described *Corallina officinalis* and a few other coralline algae. Ellis apparently had a positive influence on Linnaeus in regard to marine algae, and they occasionally corresponded. An exception is Pallas[344a] who, while placing coralline algae among the animals (possibly under pressure to do so), believed they were plants, describing *Millepora*.[187] In an important paper, Ellis and Solander[156] described several new species of articulated coralline algae (and *Halimeda*) as animals resembling plants.

The first paper of note in the 1800s was by Lamouroux[255] who placed in his family Corallineae nine genera, four of them coralline algae: *Melobesia, Corallina, Jania,* and *Amphiroa*. He followed this with a book[256] in which he described numerous species. This was an important contribution, although even Lamouroux recognized cor-

Table 1
MAJOR CONCEPTS OF THE CORALLINE ALGAE IN THE PAST TWO HUNDRED YEARS

Main Events	Times
Placed with coral animals and with some calcareous green algae (e.g., *Halimeda*)	1700s
Several genera described (e.g., *Amphiroa, Melobesia*)	Early 1800s
Recognized as plants	Early 1800s
Considered to be a single group with nonarticulated and articulated subgroups	Mid 1800s
Vegetative and reproductive structures clearly described	Late 1800s
Divided into seven subgroups, only one of which contained articulated taxa	Early 1900s
Recognized as comprising seven subgroups, with articulated taxa divided into three of the subgroups for the first time	Mid 1900s
Nonarticulated and articulated taxa placed in five subgroups, with articulated taxa in some of the same subgroups as nonarticulated taxa for the first time	Mid 1900s

alline algae as animals. The famous zoologist Lamarck[254] also included coralline algae in a treatise on corals, describing several articulated species (e.g., *Corallina ephedraea*, = *Amphiroa ephedraea*). Gray[193] clearly considered his concept of marine plants as including *Jania* and *Corallina*. Finally, Philippi[353] described *Lithothamnium* and *Lithophyllum* and emphasized that they and *Corallina* are all related plants. Hence, the plant-like nature of these organisms finally became established, even though a few subsequent papers on algae excluded coralline algae (e.g., Harvey's, *A Manual of the British Algae,* 1841).

In the middle 1800s several scientists made great progress in furthering an understanding of the diversity of coralline algae (Tables 2 and 3). Johnston,[488] Decaisne,[128-129] Kützing,[250a.b-251] Harvey,[199] Areschoug,[32] and others described and illustrated many species, and gave schemes for grouping the taxa. Areschoug[32] receives

Table 2
INFORMATION SOURCES ON TAXONOMY AND NOMENCLATURE OF NONARTICULATED CORALLINE ALGAE AT THE GENERIC LEVEL PRIOR TO FOSLIE'S PUBLICATIONS

Genera	Ref.
Melobesia, new genus	255
Nullipora, new genus	254
Tenarea, new genus	489
Lithothamnium and *Lithophyllum*, new genera	353
Spongites, new genus	251
Mastophora, new genus	128
Mastophora, made a section under *Melobesia*	129
Hapalidium and *Pheophyllum* new genera	250a
Lithocystis, new genus	199
Choreonema, new genus	370
Archaeolithothamnium, new genus	365b
Schmitziella, new genus	36

Table 3
INFORMATION SOURCES ON TAXONOMY AND NOMENCLATURE OF ARTICULATED CORALLINE ALGAE AT THE GENERIC LEVEL PRIOR TO 1900

Genera	Ref.
Corallina, new genus	280
Amphiroa and *Jania*, new genera	255
Arthrocardia, new genus	128
Haliptilon, new group under *Jania;*	129
Euamphiroa, Eurytion, and *Cheilosporum* new sections under *Amphiroa*	
Cheilosporum, elevated to genus	492
Haliptilon, elevated to genus	491
clarified generic circumscriptions	32
Lithothrix, new genus	192

credit for placing all coralline taxa into the Melobesieae (nonarticulated species) and the Corallineae (articulated species).

A few outstanding anatomical works appeared in the latter part of the 1800s. Rosanoff[364] worked with several crustose genera and Solms-Laubach[412] focused on reproductive structures. But, at the close of the century Foslie began publishing on nonarticulated coralline algae (almost exclusively), contributing at least 80 papers before his death in 1909. Before the end of the first decade of 1900 not only Foslie, but also Heydrich,[203-206] Yendo,[454-459] and Weber-van Bosse[444] (with Foslie) presented the results of extensive studies on coralline algae, and this must be considered a peak of activity in coralline taxonomy.

From 1895 to 1911 the classification of the crustose coralline algae was studied intensively by Foslie at Trondheim, Norway and Heydrich in Germany. The taxonomic viewpoints presented by these men fluctuated considerably in their successive papers, thereby resulting in some confusion as to the generic and subgeneric limits which they suggested. In addition, the earlier publications bore evidence of ill-will between Foslie and Heydrich, a fact which is probably partly responsible for some of the resulting confusion. Nevertheless, the contributions of these two men, especially Foslie, have brought to light many fundamental differences between groups of encrusting coralline algae and form a substantial base on which subsequent workers are continuing to shape a framework of classification.

After Foslie's initial publication[160] on some new species and varieties of *Lithothamnium* (as including *Lithophyllum*) Heydrich[203] published a paper in which he revised the genera of the Corallinaceae. In addition to erecting the genus *Sporolithon*, which is referred to *Archaeolithothamnium* Rothpletz[365b] by many phycologists, he recognized *Choreonema, Melobesia, Mastophora, Lithophyllum, Lithothamnium, Amphiroa, Cheilosporum,* and *Corallina*. The six nonarticulated genera were separated primarily on the basis of vegetative structures; *Sporolithon* and *Lithothamnium* were separated on the basis of tetrasporangial arrangement.

Foslie's reply[160a] is especially critical of *Sporolithon*, claiming that the sporangial cavities are probably actually cavities produced by boring animals. Heydrich[203a] replied by presenting convincing evidence of the nature of the sporangia in *Sporolithon* and erected the genus *Epilithon*, based on the type of *Melobesia, M. membranacea*. This latter move has since caused much nomenclatural confusion regarding *Melobesia* and *Epilithon*.[309]

Foslie[160a] recognized that the species named by Heydrich under *Sporolithon* should be referred to as *Archaeolithothamnium*, and made the transfers. In doing this Foslie becomes the authority for *Archaeolithothamnium*, since Rothpletz[365b-366] did not use the name in a generic sense nor did he coin any binomial for it, instead treating it under *Lithothamnium*. Thus *Sporolithon* is the earliest valid name for the group of coralline algae in which each tetrasporangium is borne within an individual cavity, a fact overlooked by most phycologists.

In the following year Foslie, perhaps hurriedly, erected five new genera of nonarticulated coralline algae, namely *Clathromorphum, Goniolithon,* and *Phymatolithon*[160c] and *Chaetolithon* and *Dermatolithon*.[160d] Despite the brief nature of the circumscriptions these genera are at present recognized by most phycologists, although some are in an amended form. Foslie had a great influence on taxonomy, and his work forms a base from which to build newer schemes as more information becomes available. Adey[13] has examined the hundreds of important specimens housed in the Foslie Herbarium in Trondheim, Norway. A scheme of classification proposed by Foslie[161b] is given in Table 4.

Yendo,[454-455,457-459] working with articulated coralline algae in Japan and British Columbia, was the first to recognize the rich diversity in Japan (Table 5). He prepared

Table 4
THE CLASSIFICATION SCHEME FOR
NONARTICULATED CORALLINE ALGAE GIVEN BY
FOSLIE[161b] AND MODIFIED BY HIM IN 1905.[162(a)]

Lithothamnioneae (tribe)	Melobesieae (tribe)
Lithothamnium (genus)	*Goniolithon* (genus)
Crustacea (group)	*Eugoniolithon* (subgenus)
Subramosa	*Hydrolithon*
Ramosa	*Melobesia*
Epilithon	*Eumelobesia* (subgenus)
Archaeolithothamnium	*Heteroderma*
Phymatolithon	*Lithophyllum*
Euphymatolithon	*Eulithophyllum* (subgenus)
Clathromorphum	Crustacea
Mastophoreae (tribe)	Ramosa
Mastophora	*Lepidomorphum* (subgenus)
Eumastophora (subgenus)	Crustacea
Lithoporella	Subramosa
	Ramosa
	Dermatolithon (subgenus)
	Carpolithon (subgenus)

ᵃ Adapted from Littler.[284]

the first descriptions of species in *Marginisporum, Alatocladia,* and *Serraticardia,* naming a total of 24 new species and varieties. He[457] produced a lengthy study on the comparative anatomy of genicula, revealing their variety for the first time (and in English). Also in 1904, Weber-van Bosse[444] (on articulates) and Foslie[161b] (on nonarticulates) published an excellently illustrated anatomical-taxonomic paper based mostly on specimens from the South Pacific. In this paper Weber-van Bosse described the genus *Metagoniolithon*.

Providing taxonomic highlights in the middle part of the present century are Lemoine,[87,269-277] on vegetative anatomy in nonarticulated taxa, Manza,[294-301] on articulated coralline algae, especially from California and South Africa, and Adey,[3-18] and Masaki[303-308] on epilithic crustose coralline algae. Also publishing currently are Cabioch,[62-77] on vegetative morphogenesis, Johansen,[221-233] on articulated taxa, and Lebednik[261-265] on nonarticulated taxa.

Others are notable for other types of contributions, such as Segawa,[377-389] with careful anatomical studies of Japanese articulated coralline algae, and Dawson,[111-121] with numerous papers on floristics and taxonomy, mostly on algae from Pacific Mexico.

RECOGNIZING THE GENERA

Almost all of the characteristics that have been employed in segregating coralline genera are structural. Two approaches have been made, one delimiting genera mostly on the basis of vegetative structures, and the other doing the same, but with reproductive structures. For example, the French phycologists Lemoine and Cabioch have used vegetative features almost exclusively. Nevertheless, there are numerous reproductive features that are excellent for understanding differences among genera and even tribes (e.g., the Corallineae and the Janieae).

Many characteristics of coralline algae are potentially useful in delimiting the genera. The tabular keys in Johansen[232] allow for the selection of taxonomically important features after a quick scan. A tabular key is in two parts, namely, a listing of the characteristics and their variants, and a table listing the taxa and the cognates identi-

Table 5
INFORMATION SOURCES ON TAXONOMY AND NOMENCLATURE OF ARTICULATED CORALLINE ALGAE AT THE GENERIC LEVEL FROM 1900 TO THE PRESENT

Genera	Ref.
Cheilosporum, broadened concept	456
Metagoniolithon and *Litharthron*, new genera	444
Marginisporum, new section under *Amphiroa*;	459
Eucheilosporum, *Serraticardia*, and *Alatocladia*, new sections under *Cheilosporum*; *Officinales*, new section under *Corallina*	
Litharthron, recognized as noncorallinaceous	493
Duthiea, *Bossea*, *Calliarthron*, *Joculator*, and *Pachyarthron*, new genera; *Cornicularia*, new subgenus under *Corallina*	300
Duthiophycus, new name for *Duthiea*	429
Joculator, merged with *Corallina*	113
Yamadaea, new genus	389
Bossiella, new name for *Bossea*; *Serraticardia*, elevated to genus	405
Chiharaea, new genus	221
Marginisporum, elevated to genus	175-177
Duthiophycus, merged with *Arthrocardia*; *Alatocladia*, elevated to genus; *Pachyarthron*, made subgenus under *Bossiella*	223

fying the appropriate variants. Three tabular keys for some of the genera in the Melobesioideae, Mastophoroideae, and Corallinoideae follow.

Characteristics, variants, and cognitive symbols for the tabular key to the genera of Melobesioideae in Table 6. *Chaetolithon* and *Antarcticophyllum,* are excluded.

1. Habitat and substrate relationships (hab).
    epi = Epilithic, or growing on hard surfaces, or epiphytic. Some epilithic species forming marl.
    par = parasitic, pigments completely or partially absent

Table 6
A TABULAR KEY GIVING MOST OF THE
GENERA IN THE MELOBESIOIDEAE
AND CHARACTERISTIC VARIANTS AS
GIVEN IN THE TEXT. ADAPTED FROM
JOHANSEN[232]

	hab 1	hyp 2	per 3	epi 4	mal 5
Clathromorphum	epi	mul	thc	2+	ovr
Kvaleya	par	uni	thc	rou	sur
Leptophytum	epi	mul	thc	rou	sur
Lithothamnium	epi	mul	thc	ang	sur
Mastophoropsis	epi	mul	thn	ang	?
Melobesia	epi	uni	thn	rou	sur
Mesophyllum	epi	cox	thc	rou	sur
Phymatolithon	epi	mul	thc	rou	sur
Synarthrophyton	epi	cox	thc	rou	sur

2. Hypothallus (hyp)
   uni = Unistratose
   mul = Multistratose; cells not arranged in conspicuous layers
   cox = Multistratose and coaxial
   × = Hypothallus absent as a distinct tissue
3. Perithallus (per)
   thc = Thick
   thn = Thin, sometimes absent except around conceptacles
   × = Perithallus absent as a distinct tissue
4. Epithallus (epi)
   ang = Mostly one layer of cells each of which is angular (eared) in section
   rou = Mostly one layer of cells each of which is rounded in section
   2+ = Two to several layers of rounded cells
5. Male conceptacles (mal)
   sur = Roof formed by overgrowing of filaments surrounding fertile area
   ovr = Roof formed by growth of filaments remaining intact over fertile areas; Chamber appears when tissue splits in a plane parallel to the surface of the crust.

Characteristics, variants and cognitive symbols for the tabular key to the genera of the Mastophoroideae in Table 7

1. Habitat and substrate relationships (hab)
   epi = Epilithic or growing on hard surfaces, or epiphytic
   par = Parasitic; all or parts of plants growing within host; pigments sometimes completely or partially absent
   rib = Ribbonlike and branched; attached to substrate at one end only
2. Hypothallus (hyp)
   uni = Unistratose
   mul = Multistratose
   × = Hypothallus absent as a distinct tissue
3. Perithallus (per)
   thc = Thick
   thn = Thin, sometimes absent except around conceptacles
   × = Perithallus absent as a distinct tissue, at least in vegetative parts

Table 7
A TABULAR KEY GIVING THE GENERA IN THE MASTOPHOROIDEAE AND CHARACTERISTIC VARIANTS AS GIVEN IN THE TEXT. ADAPTED FROM JOHANSEN[232]

	hab 1	hyp 2	per 3	meg 4
Choreonema	par	×	×	×
Fosliella	epi	uni	thn	sin
Heteroderma	epi	uni	thn	×
Hydrolithon	epi	uni	thc	ver
Litholepis	epi	uni	×	×
Lithoporella	epi	uni	×	sin
Mastophora	rib	uni	×	sin
Neogoniolithon	epi	mul	thc	ver
Porolithon	epi	mul	thc	hor

Table 8
A TABULAR KEY GIVING THE GENERA IN THE CORALLINOIDEAE AND CHARACTERISTIC VARIANTS AS GIVEN IN THE TEXT. ADAPTED FROM JOHANSEN[232]

	bra 1	med 2	pri 3	mal 4	fus 5
Alatocladia	bot	flx	a/m	con	bro
Arthrocardia	pin	str	axi	con	bro
Bossiella	bot	str	cor	con	bro
Calliarthron	bot	flx	m/c	con	bro
Cheilosporum	dic	str	mar	wal	nar
Chiharaea	bot	str	a/m	con	bro
Corallina	pin	str	axi	con	bro
Haliptilon	bot	str	axi	wal	nar
Jania	dic	str	axi	wal	nar
Marginisporum	bot	str	m/c	con	bro
Serraticardia	pin	str	a/c	con	bro
Yamadaea	rar	str	axi	con	bro

4. Megacells (meg)
   ver = Single or grouped in vertical rows in thick thalli
   hor = Grouped in horizontal rows
   sin = Single, in thin thalli
   × = Megacells absent

Characteristics, variants and cognitive symbols for the tabular key to the genera of Corallinoideae in Table 8.

1. Branching (bra)
   rar = Rarely or never branched. One genus.
   pin = Pinnate.

dic = Dichotomous.
bot = Both pinnate and dichotomous, or irregular in some species.
2. Intergenicular medulla (med)
   str = Filaments straight
   flx = Filaments flexuous.
3. Conceptacle primordia (pri)
   axi = Primordia axial only.
   mar = Marginal only.
   cor = Primordia cortical only.
   a/c = Axial in male plants and cortical in female and tetrasporangial plants. One genus.
   a/m = Axial and marginal.
   m/c = Marginal and cortical.
4. Male conceptacles (mal)
   con = Floor concave and chamber broad; fertile area on floor and extending up walls to roof; canal more than 100 μm long.
   wal = Floor concave and chamber narrow; most of fertile area lining high walls of chamber; canal less than 100 μm long
5. Fusion cells (fus)
   nar = Narrow (less than 90 μm), thick (more than 20 μm) and intact
   bro = Broad (more than 90 μm), thin (less than 20 μm) and intact or not intact.

CURRENT SCHEMES OF CLASSIFICATION

There are two main schemes of classification on the suprageneric level in vogue at the present time, one of which has been adopted in this book. The other scheme is advocated by Cabioch and places less emphasis on the presence of genicula as being an evolutionary step worthy of recognition on the subfamily level. Largely on the basis of vegetative anatomy and cytology studies, Cabioch proposed a novel scheme and placed all articulated genera in subfamilies containing nonarticulated genera. Important in delimiting her subfamilies are the presence of cellular fusions or secondary pit-connections, the single or multiple pore characteristics of asexual conceptacles, and the markedly reduced nature of *Schmitziella*. The tribes within the subfamilies were distinguished by differences in vegetative tissues and within the tribes the genera were arranged into groups containing forms that she felt (1) had stagrated evolutionarily, (2) are of an intermediate stage, and (3) show advanced characteristics. Cabioch's system, first presented in 1971 and amplified in 1972, is as follows; the subfamilies are numbered:

1. Schmitzielloideae
   Schmitzielleae: *Schmitziella*
2. Sporolithoideae
   Sporolitheae: *Sporolithon*
3. Lithothamnioideae
   Lithothamnieae: *Kvaleya, Melobesia, Lithothamnium, Polyporolithon*, and *Mesophyllum*
4. Corallinoideae
   Mastophoreae: *Lithoporella, Mastophora*, and *Metamastophora*
   Neogoniolithoneae: *Fosliella, Hydrolithon (not included in Cabioch)*[76] *Neogoniolithon, Porolithon*, and *Metagoniolithon*

Table 9
THE SCHEME OF SUPRAGENERIC CLASSIFICATION IN WHICH ARTICULATED TAXA ARE PLACED IN DIFFERENT SUBFAMILIES FROM NONARTICULATED TAXA[a]

Family Corallinaceae Lamouroux
  Subfamily Schmitzielloideae (Foslie ex Svedelius) Johansen
  Subfamily Melobesioideae (J. Areschoug) Yendo
    Tribe Lithothamnieae Foslie
    Tribe Phymatolitheae Adey et Johansen
  Subfamily Mastophoroideae Setchell
  Subfamily Lithophylloideae Setchell
  Subfamily Corallinoideae
    Tribe Corallineae
    Tribe Janieae Johansen et Silva
  Subfamily Metagoniolithoideae Johansen
  Subfamily Amphiroideae Johansen
    Tribe Amphiroeae
    Tribe Lithotricheae Johansen et Silva

[a] Adapted from Adey and Johansen[23] and Lebednik.[263]

Corallineae:	*Choreonema, Yamadaea, Chiharaea, Alatocladia, Arthrocardia, Bossiella, Calliarthron, Cheilosporum, Marginisporum, Serraticardia, Corallina, Jania,* and *Haliptilon*
5. Lithophylloideae	
Amphiroeae:	*Amphiroa*
Lithophylleae:	*Pseudolithophyllum, Lithophyllum*
Dermatolitheae:	*Dermatolithon, Tenarea, Goniolithon,* and *Lithothrix*

This scheme of classification contrasts with the views held in papers by Johansen,[223] Adey and Johansen,[23] and Adey and MacIntyre.[24] Here, as given in Tables 3 and 4 in Chapter 1, nonarticulated and articulated genera are placed in separate subfamilies. The presence of the unique genicula and the adaptations made possible by the appearance of flexible areas seem worthy of this segregation. In the latest scheme, shown in Table 9, some of the subfamilies contain tribes. The most recent recognition of tribes has been in two articulated subfamilies Corallinoideae and Amphiroideae.[236] This attempt to more naturally reflect phylogenetic tendencies in these two subfamilies is supported by several structural differences as shown in Tables 10 and 11.

Finally, the uses of modern techniques may provide additional information that can be used in systematics. In one of the first uses of scanning electron microscopy of coralline surfaces, Garbary[181] categorized 16 species in 9 genera on the structure and configuration of epithallial cell "concavities" and other surface features. In a dichotomous key he used features such as: concavities in rows separated by grooves or ridges, calcification heavier on one side of each concavity, concavities on conceptacle roof cup-shaped or flat, relative a real coverage of concavities, and even concavities lacking (see Chapter 2). This technique has great promise, although these preliminary results show that some characteristics are shared by unrelated genera, such as cup-shaped concavities on conceptacle roofs in *Dermatolithon litorale* and *Fosliella minutula*.

ADAPTATIONS IN CORALLINE ALGAE

It is interesting to speculate on the adaptive significance of lime deposition in coral-

Table 10
CHARACTERISTICS DISTINGUISHING CORALLINEAE FROM
JANIEAE. ADAPTED FROM JOHANSEN AND SILVA[236]

Characteristics	Corallineae	Janieae
Fusion cells and gonimoblast filaments	Thin (<12 μm), expanded (90—300 μm broad); filaments arising from margins and/or upper surfaces	Thick (up to 35 μm), compact (40—130 μm) broad; filaments arising from margins only
Male conceptacles	Chamber broad (350—450 μm); canal long (200—500 μm or more)	Chamber narrow (90—250 μm); canal short (30—120 μm)
Number of tetrasporangia per conceptacle	many, more than 30	Few, <15
Number of supporting cells per procarpic conceptacle	Many, >200	Few, <200
Number of tetrasporangial initials per conceptacle	Many, >200	Few, <200
Tetraspores	Relatively small,	Relatively large
Spore germination and resulting crust	slow; crust extensive	rapid; crust reduced
Branching of main axis	Pinnate or dichotomous	Dichotomous
Length of intergenicular medullary cells	Short (50-90 μm)	Long (80—170 μm)
infection by *Choreonema*	No	Yes

Table 11
CHARACTERISTICS DISTINGUISHING AMPHIROEAE FROM
LITHOTRICHEAE. ADAPTED FROM JOHANSEN AND SILVA[236]

Characteristics	Amphiroeae	Lithotricheae
Number of tiers of medullary cells in genicula	One or (in most species) several; cells becoming decalcified but not elongating markedly	One; cells elongating to as much as 600 μm
Number of tiers of medullary cells in intergenicula	Several	One
Ration of length of genicular cells to intergenicular cells	1—2	As much as 40
Origin of cortex	From both genicula and intergenicula	From intergenicula only
Thickness of cortex	Several cells thick	Unistratose (sometimes proliferating with age)
Calcified cortex covering Genicula	None, or cracking away irregularly	Intergenicular cortex growing down over genicula
Crustose base	Cells nearly isodiametric	Cells elongate

line algae. The hardening suggests that resistance to grazing, wave shock, or abrasion, may have contributed to the success of coralline algae.[288] Although fragments of coralline algae have been reported from the guts of grazing animals, it seems unlikely that coralline thalli are an important food for any animals. Many coralline algae grow in areas that are under extreme stress from moving water. They seem successful in this kind of environment, but so do some noncalcified marine algae. Abrasion is not a feature encountered in the ocean except on seashores frequented by man. It is possible that resistance to growth by fouling epiphytes is greater in coralline algae than in noncalcareous plants. As mentioned earlier, the sloughing of outer acellular and cellular

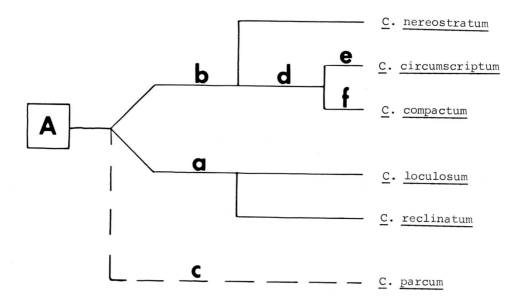

FIGURE 1. Diagram showing possible evolutionary routes resulting in the presently recognized species of *Clathromorphum*. The hypothetical ancestor (A) had diffuse conceptacles, heavy cuticle, 1-2 epithallial cell layers, no branching lines, and a large-celled hypothallus with ascending and descending filaments. a = perithallial cells large; b = perithallial cells small; c = no marginal meristem, unique end walls; d = arctic distribution, bispores, reduction of hypothallus; e = deciduous conceptacle patches; f = no patches. Adapted from Lebednik.[262]

layers reduces attack by epiphytes. Littler[288] mentioned that protection against intense light could be a factor of adaptive importance leading to the evolution of calcified cell walls. While certain species, such as those in *Porolithon,* are able to withstand high light intensities, many coralline algae grow best at low light intensities.

Speculation may also be made about possible advantages that branched fronds with uncalcified genicula have over nonarticulated forms. Two advantages of flexibility seem possible. The flexibility afforded by genicula allows for longer branches that can extend upward without being damaged by the mechanical pressure of water and organisms. This greater elevation would enable the plants to compete more successfully for light. A second possibility is that wider spore dispersal is realized when they are shed farther above the sea floor.

A consideration of *Yamadaea* and *Chiharaea* allows this speculation to go one step more. The flattened, recumbent intergenicula of *Chiharaea* offer less resistance to surging waves than do those of other more erect and bushy articulated coralline algae. The location of conceptacular pores on the upper surfaces rather than at the apices of the prostrate intergenicula may facilitate spore discharge into water currents. It seems reasonable, therefore, to consider that *Chiharaea* evolved from larger segmented ancestors with deeply embedded conceptacles which were subjected to water stresses such as in the habitats where *Chiharaea bodegensis* occurs.

In several parts of this book allusions have been made to possible evolutionary paths taken by coralline algae. When intensive studies are made of a subgroup of the Corallinaceae it is feasible to construct a phylogeny, such as Lebednik[262] did for *Clathromorphum* (Figure 1). There are certain to be more studies in which possible phylogenetic routes are presented. To do this is most tempting and, with the current proliferation of information on coralline algae, it will become impossible to resist. Although there are some glaring holes in our taxonomic knowledge of coralline algae (e.g., *Chaetolithon*), in some respects these plants are now among the best known seaweeds.

SUMMARY

A great number of papers provide structural information that is useful in understanding the taxonomy of coralline algae. It is necessary to unscramble much old nomenclature produced in the last two centuries. Variability in form must be considered in working on coralline biosystematics.

Coralline algae were treated as lime-producing animals until their plant-like nature was revealed by anatomical studies in the 1800s. For convenience, they are divided into nonarticulated and articulated species, but now most students of the group recognize five to seven subfamilies in the family Corallinaceae. The organization of genera into subfamilies has an extremely complex history, with some of the more prominent scientists involved included Lamouroux, Foslie, Decaisne, Areschoug, Kützing, and Yendo.

Even though generic concepts are still in a state of flux, it is possible to construct tabular keys and examine the main distinguishing features relatively quickly. Tabular keys are given for the three largest subfamilies.

Two main schemes for grouping coralline algae into subfamilies and tribes are available. In one, articulated genera are placed in the same subfamilies with nonarticulated genera. Emphasis is placed on simple versus complex thallus structure. In the other scheme genicula are considered important and the nonarticulated and articulated genera are placed in separate subfamilies. This scheme, with seven subfamilies, is used in this book.

The success of coralline algae suggests that features such as calcium carbonate, conceptacles and epithallia are desirable in marine environments. Intensive studies of certain groups, for example *Clathromorphum,* allows for speculation on phylogeny. However, there are some areas and taxa where there is hardly any information at all, for example, *Chaetolithon.*

REFERENCES

1. **Abbott, I. A. and Hollenberg, G. J.**, *Marine Algae of California,* Stanford University Press, Calif., 1976.
2. **Abbott, I. A. and North, W. J.**, Temperature influences on floral composition in California coastal waters, in *Proc. 7th Int. Seaweed Symp.* Nizizawa, K., Ed., University of Tokyo Press, Tokyo, 1972, 72.
3. **Adey, W. H.**, The genus *Phymatolithon* in the Gulf of Maine, *Hydrobiologia,* 24, 377, 1964.
4. **Adey, W. H.**, The algal tribe Lithophylleae and its included genera. II, Q. Colo. Sch. Mines, 60, 67, 1965.
5. **Adey, W. H.**, The genus *Clathromorphum* (Corallinaceae) in the Gulf of Maine, *Hydrobiologia,* 26, 539, 1965.
6. **Adey, W. H.**, Distribution of saxicolous crustose corallines in the northwestern North Atlantic, *J. Phycol.,* 2, 49, 1966.
7. **Adey, W. H.**, The genus *Pseudolithophyllum* (Corallinaceae in the Gulf of Maine), *Hydrobiologia,* 27, 479, 1966.
8. **Adey, W. H.**, The genera *Lithothamnium, Leptophytum* (nov. gen.), and *Phymatolithon* in the Gulf of Maine, *Hydrobiologia,* 28, 321, 1966.
9. **Adey, W. H.**, The distribution of crustose corallines on the Icelandic coast, *Sci. Islandica,* 1968, 16, 1968.
10. **Adey, W. H.**, Some relationships between crustose corallines and their substrate, *Sci. Islandica,* 2, 21, 1970.
11. **Adey, W. H.**, The crustose corallines of the northwestern North Atlantic, including *Lithothamnium lemoineae* n. sp., *J. Phycol.,* 6, 225, 1970.
12. **Adey, W. H.**, The effects of light and temperature on growth rates in boreal-subarctic crustose corallines, *J. Phycol.,* 6, 269, 1970.
13. **Adey, W. H.**, A revision of the Foslie crustose coralline herbarium, *K. Nor. Vidensk. Selsk. Skr.,* 1970(1), 1, 1970.
14. **Adey, W. H.**, The sublittoral distribution of crustose corallines on the Norwegian coast, *Sarsia,* 46, 41, 1971.
15. **Adey, W. H.**, Temperature control of reproduction and productivity in a subarctic coralline alga, *Phycologia,* 12, 111, 1973.
16. **Adey, W. H.**, The algal ridges and coral reefs at St. Croix: their structure and Holocene development, *Atoll Res. Bull.,* 187, 1975.
17. **Adey, W. H.**, Shallow water Holocene bioherms of the Caribbean Sea and West Indies, Proc. 3rd Coral Reef Symp. 2. University of Miami, Fla., 1977.
18. **Adey, W. H.**, Coral reef morphogenesis: a multidimensional model, *Science,* 202, 831, 1978.
19. **Adey, W. H.**, Algal ridges of the Caribbean Sea and West Indies, *Phycologia,* 17, 361, 1978.
20. **Adey, W. H. and Adey, P. J.**, Studies on the biosystematics and ecology of the epilithic crustose Corallinaceae of the British Isles, *Br. Phycol. J.,* 8, 343, 1973.
21. **Adey, W. H., Adey, P. J., Burke, R., and Kaufman, L.**, The Holocene reef systems of Eastern Martinique, French West Indies, *Atoll Res. Bull.,* 218, 1977.
22. **Adey, W. H. and Burke, R.**, Holocene bioherms (algal ridges and bank-barrier reefs) of the eastern Caribbean, *Geol. Soc. Am. Bull.,* 87, 95, 1976.
23. **Adey, W. H. and Johansen, H. W.**, Morphology and taxonomy of Corallinaceae with special reference to *Clathromorphum, Mesophyllum,* and *Neopolyporolithon* gen. nov. (Rhodophyceae, Cryptonemiales), *Phycologia,* 11, 159, 1972.
24. **Adey, W. H. and MacIntyre, I. G.**, Crustose coralline algae: a reevaluation in the geological sciences, *Geol. Soc. Am. Bull.,* 84, 883, 1973.
25. **Adey, W. H., Masaki, T., and Akioka, H.**, *Ezo epiyessoense,* a new parasitic genus and species of Corallinaceae (Rhodophyta, Cryptonemiales), *Phycologia,* 13, 329, 1974.
26. **Adey, W. H., Masaki, T., and Akioka, H.**, The distribution of crustose corallines in Eastern Hokkaido and the biogeographic relationships of the flora, *Bull. Fac. Fish., Hokkaido Univ.,* 26, 303, 1976.
27. **Adey, W. H., and McKibbin, D. L.**, Studies on the Maerl species *Phymatolithon calcareum* (Pallas) nov. comb. and *Lithothamnium coralloides* Crouan in the Ria de Vigo, *Bot. Mar.,* 13, 100, 1970.
28. **Adey, W. H., and Sperapani, C. P.**, The biology of *Kvaleya epilaeve,* a new parasitic genus and species of Corallinaceae, *Phycologia,* 10, 29, 1971.
29. **Adey, W. H., and Vassar, J. M.**, Colonization, succession and growth rates of tropical crustose coralline algae (Rhodophyta, Cryptonemiales), *Phycologia,* 14, 55, 1975.

30. **Alexandersson, T.**, Carbonate cementation in recent coralline algal constructions, in *Fossil Algae, Recent Results and Developments*, Flügel, E., Ed., Springer-Verlag, Berlin, 1977, 261.
31. **Ardré, F.**, Contribution a l'étude les algues marines du Portugal. I. Flore. *Port. Acta Biol., Ser (B)*, 10, 1970.
32. **Areschoug, J. E.**, Ordo XII. Corallineae, in *Species Genera et Ordines Algarum*, Vol 2, Agardh, J. G., Gleerup, Lund, 1852, 506.
33. **Baas-Becking, L. G. M. and Galliher, E. W.**, Wall structure and mineralization in coralline algae, *J. Phys. Chem.* 35, 267, 1931.
34. **Bailey, A. and Bisalputra, T.**, A preliminary account of the application of thin-sectioning, freeze-etching, and scanning electron microscopy to the study of coralline algae, *Phycologia*, 9, 83, 1970.
35. **Balakrishnan, M. W.**, The morphology and cytology of *Melobesia farinosa* Lamour. *J. Indian Bot. Soc.*, (Iyengar Commemorative Volume), 305, 1947.
36. **Batters, E. A. L.**, On *Schmitziella;* a new genus of endophytic algae, belonging to the order Corallinaceae, *Ann. Bot. (London),* 6, 185, 1892.
37. **Bauch, R.**, Die Entwicklung der Bisporen der Corallinaceen, *Planta,* 26, 265, 1937.
38. **Blanc, J. J. and Molimer, R.**, Les formations organogenes construites superficielles en Mediterranee occidentale, *Bull. Inst. Oceanogr. Fish.* 1067, 1, 1955.
39. **Blinks, L. R.**, The effect of pH upon the photosynthesis of littoral marine algae, *Protoplasm,* 57, 126, 1963.
40. **Blunden, G., Binns, W. W., and Perks, F.**, Commercial collection and utilization of Maërl, *Econ. Bot.*, 29, 140, 1975.
41. **Børgesen, F.**, The marine algae of the Danish West Indies. III. Rhodophyceae, *Dan. Bot. Ark.* 3, 145, 1917.
42. **Børgesen, F.**, Some marine algae from Mauritius. III. Rhodophyceae. Part 2. Gelidiales, Cryptonemiales, Gigartinales, *K. Dan. Vidensk. Selsk. Biol. Meddel.*, 19, 1943.
43. **Boger, E. A. and Johansen, H. W.**, Plastiquinones in coralline algae (Corallinaceae), *Phyton*, 32, 129, 1974.
44. **Bonar, L.**, Studies in some California fungi IV, *Mycologia,* 57, 379, 1965.
45. **Borowitzka, M. A.**, Algal calcification, *Oceanogr. Mar. Biol.*, 15, 189, 1977.
46. **Borowitzka, M. A.**, Plastid development and floridean starch grain formation during carposporogenesis in the coralline red alga *Lithothrix aspergillum* Gray, *Protoplasma*, 95, 217, 1978.
47. **Borowitzka, M. A.**, Calcium exchange and the measurement of calcification in the calcareous coralline red alga *Amphiroa foliacea.*, *Mar. Biol.*, 50, 339, 1979.
48. **Borowitzka, M. A., Larkum, A. W. D., and Borowitzka, L. J.**, A preliminary study of algal turf communities of a shallow coral reef lagoon using an artificial substratum, *Aquatic Bot.*, 5, 365, 1978.
49. **Borowitzka, M. A., Larkum, A. W. D., and Nockolds, C. E.**, A scanning electron microscope study of the structure and organization of the calcium carbonate deposits of algae, *Phycologia,* 13, 195, 1974.
50. **Borowitzka, M. A. and Vesk, M.**, Ultrastructure of the Corallinaceae. I. The vegetative cells of *Corallina officinalis* and *C. cuvierii, Mar. Biol.,* 46, 295, 1978.
50a. **Borowitzka, M. A. and Vesk, M.**, Ultrastructure of the Corallinaceae (Rhodophyta). II. Vegetative cells of *Lithothrix aspergillum, J. Phycol.,* 15, 146, 1979.
51. **Bosellini, A. and Ginsburg, R. M.**, Form and internal structure of recent algal nodules (rhodolites) from Bermuda, *J. Geol.,* 79, 669, 1971.
52. **Bosence, D.**, Ecological studies in two carbonate sediment-producing algae, in *Fossil Algae, Recent Results and Developments,* Flügel, E., Ed., Springer-Verlag, Berlin, 1977, 270.
53. **Boudouresque, C. F. and Cinelli, F.**, Sur un *Fosliella* du Golfe de Naples (Italie), *Pubbl. St. Zool. Napoli,* 39, 108, 1974.
54. **Boulanger, D. and Poignant, A. F.**, Les Nodules Algaires du Miocene D'Aquitaine Meridionale, *Bull. Cent. Etud. Rech. Sci.*, Biarritz, 10, 685, 1975.
55. **Bressan, G.**, *Litholepis mediterranea* Foslie observee a Ustica (Sicile), *Atti Ist. Veneto Sci. Lett. Arti., Cl. Sci. Mat. Nat.,* 78, 265, 1970.
56. **Bressan, G.**, Rodoficee calcaree dei mari Italiani, *Boll. Soc. Adriat. Sci., Trieste,* 59, 1, 1974.
57. **Bressan, G., Minati-Radin, D., and Smundin, L.**, Ricerche sul genere *Fosliella* (Corallinaceae — Rhodophyta): *Fosliella cruciata* sp. nov., *G. Bot. Ital.,* 111, 27, 1977.
58. **Brown, V., Ducker, S. C., and Rowan, K. S.**, The effect of orthophosphate concentration on the growth of articulated coralline algae (Rhodophyta), *Phycologia,* 16, 125, 1977.
59. **Buchbinder, B.**, The coralline algae from the Miocene Ziqlag Formation in Israel and their environmental significance, in *Fossil Algae, Recent Results and Developments,* Flügel, E., Ed., Springer-Verlag, Berlin, 1977, 279.
60. **Buge, E., Debourle, A., and Deloffre, R.**, Gisement miocene a nodules algaires (Rhodolithes) a l'Ouest de Salies-de-Bearn (Aquaitaine Sud-Ouest), *Bull. Cent. Rech. Pau,* 7, 1, 1973.

61. **Burgner, R. L., Isakson, J. S., and Lebednik, P. A.**, Observations on the effect of the Milrow Nuclear Test on marine organisms at Amchitka, *Bioscience*, 21, 671, 1971.
62. **Cabioch, J.**, Sur le mode de développement des spores chez les Corallinacées, *C. R. Acad. Sci., Ser. D*, 262D, 2025, 1966.
63. **Cabioch, J.**, Sur le mode de formation du thale articulé chez quelques Corallinées, *C. R. Acad. Sci, Ser. D*, 262D, 339, 1966.
64. **Cabioch, J.**, Contribution à l'étude morphologique, anatomique et systématique de deux Mélobésiees: *Lithothamnium calcareum* (Pallas) Areschoug et *Lithothamnium corallioides* Crouan, *Bot. Mar.*, 9, 33, 1966.
65. **Cabioch, J.**, Sur le mode de formation des triochocytes chez le *Neogoniolithon notarisii* (Dufour) Setchell et Mason, *C. R. Acad. Sci., Ser. D*, 266D, 333, 1968.
66. **Cabioch, J.**, Quelques particularités anatomiques du *Lithophyllum fasciculatum* (Lamarck) Foslie, *Bull. Soc. Bot. Fr.*, 115, 173, 1968.
67. **Cabioch, J.**, Persistance de stades juvéniles et possibilité d'une néotenie chez le *Lithophyllum incrustans* Philippi, *C. R. Acad. Sci., Ser. D.*, 268D, 497, 1969.
68. **Cabioch, J.**, Les fonds de maërl de la baie du Morlaix et leur peuplement végétal, *Cah. Biol. Mar.*, 10, 139, 1969.
69. **Cabioch, J.**, Sur le mode de développement de quelques *Amphiroa* (Rhodophycées, Corallinacées), *C. R. Acad. Sci., Ser. D*, 269D, 2238, 1969.
70. **Cabioch, J.**, Le'origine des mégacytes chez les *Porolithon* (Corallinacées, Rhodophycées), *C. R. Acad. Sci., Ser. D*, 270D, 474, 1970.
71. **Cabioch, J.**, Application des caractères morphogénétiques à la systématique des Corallinacées: le genre *Goniolithon*, *C. R. Acad. Sci., Ser. D*, 270D, 1447, 1970.
72. **Cabioch, J.**, Sur l'importance des phénomènes cytologiques pour la systématique et la phylogénie des Corallinacées (Rhodophycées, Cryptonémiales), *C. R. Acad. Sci., Ser. D*, 271D, 296, 1970.
73. **Cabioch, J.**, Le maërl des cotes de Bretagne et la problème de sa survie, *Penn Bed*, 7, 421, 1970.
74. **Cabioch, J.**, Essai d'une nouvelle classification des Corallinacées actuelles, *C. R. Acad. Sci., Ser. D*, 272D, 1616, 1971.
75. **Cabioch, J.**, Étude sur les Corallinacées. I. Caractères généraux de la cytologie, *Cah. Biol. Mar.*, 12, 121, 1971.
76. **Cabioch, J.**, Étude sur les Corallinacées. II. La morphogenèse; conséquences systématiques et phylogénétiques, *Cah. Biol. Mar.*, 13, 137, 1972.
77. **Cabioch, J.**, Un fond de maerl d l'Archipel de Madère et son peuplement végétal, *Soc. Phycol. Fr., Bull.*, 19, 74, 1974.
78. **Cabioch, J. and Giraud, G.**, Comportement cellulaire au cours de la régénération directe chez de Mesophyllum lichenoides (Ellis) Lemoine (Rhodophycées, Corallinacées), *C. R. Acad. Sci., Ser. D*, 286D, 1783, 1978.
79. **Cabioch, J. and Giraud, G.**, Apport de la microscopie électronique à la comparaison de quelques especès de *Lithothamnium* Philippi, *Phycologia*, 17, 369, 1978.
80. **Cardinal, A.**, Inventaire des algues marines benthiques de la Baie des Chaleurs et de la Baie de Gaspe (Quebec). III. Rhodophycees, *Nat. Can.*, 94, 735, 1967.
81. **Cardinal, A. and Cabioch, J.**, Les corallinacees (Rhodophyta, Cryptonemiales) des cotes du Quebec, *Cah. Biol. Mar.*, 19, 175, 1978.
82. **Celan, M. and Bavaru, A.**, Contribution a la connaissance des algues rouges (Rhodophycees) de la mer Noire, *Rev. Roum. Biol., Ser. Bot.*, 12, 345, 1967.
83. **Chamberlain, Y. M.**, Marine algae of Gough Island, *Bull. Br. Mus. (Nat. Hist.), Bot.*, 3, 173, 1965.
84. **Chamberlain, Y. M.**, The occurrence of *Fosliella limitata* (Foslie) Ganesan (a new British Record) and *F. lejolisii* (Rosanoff) Howe (Rhodophyta, Corallinaceae) on the Isle of Wight, *Br. Phycol. J.*, 12, 67, 1977.
85. **Chamberlain, Y. M.**, Observations on *Fosliella farinosa* (Lamour.) Howe (Rhodophyta, Corallinaceae) in the British Isles, *Br. Phycol. J.*, 12, 343, 1977.
86. **Chamberlain, Y. M.**, Investigation of taxonomic relationships amongst epiphytic, crustose Corallinaceae, in *Modern Approaches to the Taxonomy of Red and Brown Algae*, Systematics Association Special Vol. 10, Irvine, D. E. G. and Price, J. H., Eds., Academic Press, New York, 1978, 223.
87. **Chamberlain, Y. M.**, A tribute to Mme. Marie Lemoine, *Phycologia*, 17, 359, 1978.
88. **Chamberlain, Y. M.**, *Dermatolithon litorale* (Suneson) Hamel and Lemoine (Rhodophyta Corallinaceae) in the British Isles, *Phycologia*, 17, 396, 1978.
89. **Chapman, V. J.**, Notes on New Zealand algae, *Trans. R. Soc. N. Z.*, 79, 84, 1951.
90. **Chave, K. E.**, Carbonates: association with organic matter in surface seawater, *Science*, 148, 1723, 1965.
91. **Chave, K. E. and Wheeler, E. D., Jr.**, Mineralogic changes during growth in the red alga, *Clathromorphum compactum*, *Science*, 147, 621, 1965.

92. **Chavez, M. L.,** Unz nueva especie de corallinacea: *Jania huertae* (Rhodoph., Florid.), *Ciencia, (Mexico City)*, 27, 133, 1972.
93. **Chihara, M.,** Recent studies on the systematics of coralline algae, *Bull. Jpn. Soc. Phycol.*, 17, 113, 1969.
94. **Chihara, M.,** Reproductive cycles and spore germination of the Corallinaceae and their possible relevance in the systematics. I. *Amphiroa, Marginisporum* and *Lithothrix, J. Jpn. Bot.*, 47, 239, 1972.
95. **Chihara, M.,** Reproductive cycles and spore germination in the Corallinaceae and their possible relevance in the systematics. II. Serraticardia and the related genera, *J. Jpn. Bot.*, 47, 306, 1972.
96. **Chihara, M.,** Reproductive cycles and spore germination of the Corallinaceae and their possible relevance in the systematics. III. *Corallina, Jania* and their related genera, *J. Jpn. Bot.*, 48, 13, 1973.
97. **Chihara, M.,** The significance of reproductive and spore germination characteristics in the systematics of the Corallinaceae: articulated coralline algae, *Jpn. J. Bot.*, 20, 369, 1973.
98. **Chihara, M.,** Reproductive cycles and spore germination of the Corallinaceae and their possible relevance in the systematics. IV. *Lithophyllum, Lithothamnium* and their related genera, *J. Jpn. Bot.*, 48, 345, 1973.
99. **Chihara, M.,** Reproductive cycles and spore germination of the Corallinaceae and their possible relevance in the systematics. V. Five species of *Fosliella, J. Jpn. Bot.*, 49, 89, 1974.
100. **Chihara, M.,** The significance of reproductive and spore germination characteristics to the systematics of the Corallinaceae: non-articulated coralline algae, *J. Phycol.*, 10, 266, 1974.
101. **Chuvashov, B. I.,** *Katavella*, a new genus of fossil red algae, *Paleont. zh.*, 2, 144, 1965, (trans.) *Int. Geol. Rev.*, 8, 89, 1966.
102. **Colthart, B. J. and Johansen, H. W.,** Growth rates of *Corallina officinalis* (Rhodophyta) at different temperatures, *Mar. Biol.*, 18, 46, 1973.
103. **Connor, J. and Adey, W.,** The benthic algal composition, standing crop, and productivity of a Caribbean algal ridge, *Atoll Res. Bull.*, 211, 1977.
104. **Conover, J. T.,** Seasonal growth of benthic marine plants as related to environmental factors in an estuary, *Inst. Mar. Sci.*, 5, 97, 1958.
105. **Conover, J. T.,** The ecology, seasonal periodicity, and distribution of benthic plants in some Texas lagoons, *Bot. Mar.*, 7, 4, 1964.
106. **Cotton, A. D.,** *Lithophyllum* in the British Isles, *J. Bot., (London)*, 49, 115, 1911.
107. **Darley, W. W.,** Silification and calcification, in *Algal Physiology and Biochemistry*, Stewart, W. D. P., Ed., University of California Press, Berkeley, 1974, 655.
108. **Davies, P. J. and Kinsey, D. W.,** Holocene reef growth — One Tree Island, Great Barrier Reef, *Mar. Geol.*, 24, M1, 1977.
109. **Davis, B. M.,** Kerntheilung in der Tetrasporenmutterzelle bei *Corallina officinalis* L. var. *mediterranea, Ber. Dtsch. Bot. Ges.*, 16, 266, 1898.
109a. **Dawson, E. Y.,** The marine algae of the Gulf of California, *Allan Hancock Pac. Exped.*, 3, 189, 1944.
110. **Dawson, E. Y.,** Contributions toward a marine flora of the southern California Channel Islands. I-III, *Allan Hancock Found. Publ. Occasional Paper*, 8, 1, 1949.
111. **Dawson, E. Y.,** A further study of upwelling and associated vegetation along Pacific Baja California, Mexico, *J. Mar. Res.*, 10, 39, 1951.
112. **Dawson, E. Y.,** Circulation within Bahia Vizcaino, Baja California, and its effects on the marine vegetation, *Am. J. Bot.*, 39, 425, 1952.
113. **Dawson, E. Y.,** Marine red algae of Pacific Mexico. I. Bangiales to Corallinaceae subf. Corallinoideae, *Allan Hancock Pac. Exped.* 17, 1, 1953.
114. **Dawson, E. Y.,** A preliminary working key to the living species of *Dermatolithon*, in *Essays in the Natural Sciences in Honor in Captain Allan Hancock*, University of Southern California Press, Los Angeles, 1955, 271.
115. **Dawson, E. Y.,** Notes on Pacific coast marine algae VII, *Bull. South. Calif. Acad. Sci.*, 57, 65, 1958.
116. **Dawson, E. Y.,** New records of marine algae from Pacific Mexico and Central America, *Pac. Nat.*, 1, 31, 1960.
117. **Dawson, E. Y.,** Marine red algae of Pacific Mexico. III. Crytonemiales, Corallinaceae subf. Melobesioideae, *Pac. Nat.*, 2, 1, 1960.
118. **Dawson, E. Y.,** The rim of the reef, *Nat. Hist.*, 70, 8, 1961.
119. **Dawson, E. Y.,** New records of marine algae from the Galapagos Islands, *Pac. Nat.*, 4, 3, 1963.
120. **Dawson, E. Y.,** A review of Yendo's jointed coralline algae of Port Renfrew, Vancouver Island, *Nova Hedwigia, Z. Kryptogamenkd*, 7, 537, 1964.
121. **Dawson, E. Y.,** Intertidal algae, in *An Oceanographic and Biological Survey of the Southern California Mainland Shelf*, California State Water Quality Control Board Publication, Sacramento, 27, 220, 1965.

122. **Dawson, E. Y., Acleto, C., and Foldvik, N.**, The seaweeds of Peru, *Beih. Nova Hedwigia*, 13, 1, 1964.
123. **Dawson, E. Y., Neushul, M., and Wildman, R D.**, Seaweeds associated with kelp beds along southern California and northwestern Mexico, *Pac. Nat.*, 1, 1, 1960.
124. **Dawson, E. Y. and Silva, P. C.**, *Bossiella*, in *Marine Red Algae of Pacific Mexico.* I, Allan Hancock Pac. Exped., 17, 1953, 150.
125. **Dawson, E. Y. and Steele, R. L.**, An eastern Pacific member of *Yamadaia* (Corallinaceae) from the San Juan Islands, Washington, *Nova Hedwigia Z. Kryptogamenkd*, 8, 1, 1964.
126. **Dayton, P. K.**, Competition, disturbance, and community organization: the provision and subsequent utilization of space in a rocky intertidal community, *Ecol. Monogr.*, 41, 351, 1971.
127. **Dayton, P. K.**, Experimental evaluation of ecological dominance in a rocky intertidal algal community, *Ecol. Monogr.*, 45, 137, 1975.
128. **Decaisne, J.**, Essais sur une classification des algues et des polypiers calcifères de Lamouroux, *Ann. Sci. Nat. Bot.*, Ser. 2, 17, 297, 1842.
129. **Decaisne, J.**, Mémoire sur les corallines ou polypiers calcifères, *Ann. Sci. Nat. Bot.*, Ser. 2, 18, 96, 1842.
130. **Dellow, V.**, Marine algal ecology of the Hauraki Gulf, New Zealand, *Trans. R. Soc. N.Z.*, 83, 1, 1955.
131. **Dellow, V. and Cassie, R. M.**, Littoral zonation in two caves in the Aukland district, *Trans. R. Soc. N.Z.*, 83, 321, 1955.
132. **Denizot, M.**, Les algues Floridées encroutantes (a l'exclusion des Corallinacées), Thesis, Mus. Nat. Hist. Nat., Paris, (published privately by author), 1968, 310.
133. **Desikachary, T. V. and Ganesan, E. K.**, Notes on Indian red algae. IV. *Hydrolithon reinboldii* (Weber Van Bosse et Foslie) Foslie and *Hydrolithon iyengarii* sp. nov., *Phykos.*, 5, 83, 1967.
134. **Digby, P. S. B.**, Growth and calcification in the coralline algae, *Clathromorphum circumscriptum* and *Corallina officinalis* and the significance of pH in relation to precipitation, *J. Mar. Biol. Assoc. U. K.*, 57, 1095, 1977.
134a. **Digby, P. S. B.**, Photosynthesis and respiration in the coralline algae, *Clathromorphum circumscriptum* and *Corallina officinalis* and the metabolic basis of calcification, *J. Mar. Biol. Assoc., U. K.*, 57, 1111, 1977.
134b. **Digby, P. S. B.**, Reducing activity and the formation of base in the coralline algae: an electrochemical model, *J. Mar. Biol. Assoc., U. K.*, 59, 455, 1979.
135. **Dixon, P. S.**, The Rhodophyceae, in *The Chromosomes of the Algae*, Godward, M. B. E., Ed., Edward Arnold, London, 1966, 168.
136. **Dixon, P. S.**, *The Biology of the Rhodophyta*, Hafner Press, Macmillan, New York, 1973.
137. **Dommasnes, A.**, Variations in the meiofauna of *Corallina officinalis* L. with wave exposure, *Sarsia*, 36, 117, 1968.
138. **Dommasnes, A.**, On the fauna of *Corallina officinalis* L. in western Norway, *Sarsia*, 38, 71, 1969.
139. **Doty, M. S.**, Critical tide factors that are correlated with the vertical distribution of marine algae and other organisms along the pacific coast, *Ecology*, 27, 315, 1946.
140. **Doty, M. S.**, Pioneer intertidal population and the related general vertical distribution of marine algae in Hawaii, *Blumea*, 15, 95, 1967.
141. **Doty, M. S.**, Coral reef roles played by free-living algae, in *Proc. 2nd Int. Coral Reef Symp.* I, Brisbane, 1974, 27.
142. **Doty, M. S., Gilbert, W. J., and Abbott, I. A.**, Hawaiian marine algae from seaward of the algal ridge, *Phycologia*, 13, 345, 1974.
143. **Doty, M. S. and Morrison, J. P. E.**, Interrelationships of organisms on Raroia aside from man, *Atoll. Res. Bull.* 35, 1954.
144. **Ducker, S. C.**, The genus *Metagoniolithon* Weber-van Bosse (Corallinaceae, Rhodophyta), *Aust. J. Bot.*, 27, 67, 1979.
145. **Ducker, S. C., Foord, N. J., and Bruce, R. B.**, Biology of Australian Seagrasses: the genus *Amphibolis* C. Agardh (Cymodoceaceae)., *Aust. J. Bot.*, 25, 67, 1977.
146. **Ducker, S. C. and Knox, R. G.**, Alleloparasitism between a seagrass and algae, *Naturwissenschaften*, 65, 391, 1978.
147. **Ducker, S. C., LeBlanc, J. D., and Johansen, H. W.**, An epiphytic species of *Jania* (Corallinaceae: Rhodophyta) endemic to southern Australia, *Contrib. Herb. Aust.*, 17, 1, 1976.
148. **Duckett, J. G. and Peel, M. C.**, The role of transmission electron microscopy in elucidating the taxomomy and phylogeny of the Rhodophyta, in *Modern Approaches to the Taxomomy of Red and Brown Algae*, Systematics Association Special Vol. 10, Irvine, D. E. G. and Price, J. H., Eds., Academic Press, New York, 1978, 157.
149. **Duerden, R. C. and Jones, W. E.**, The host specificity of *Jania rubens* (L.) Lamour. in British Waters, *8th Int. Seaweed Symp.*, Bangor, North Wales, 1974, A36.

150. **Elliott, G. F.**, *Subterraniphyllum*, a new tertiary calcareous alga, *Paleontology*, 1, 73, 1957.
151. **Elliott, G. F.**, *Pseudaesolithon*, a calacareous alga from the Fars (Persian Miocene), *Geol. Romana*, 9, 31, 1970.
152. **Elliott, G. F.**, *Cretacicrusta* gen. nov., a possible alga from the English Cretaceous, *Palaeontology*, 15, 501, 1972.
153. **Elliott, G. F.**, A Miocene solenoporoid alga showing reproductive structures, *Palaeontology*, 16, 223, 1973.
154. **Elliott, G. F.**, Transported algae as indicators of different marine habitats in the English middli Jurassic, *Palaeontology*, 18, 351, 1975.
155. **Ellis, J.**, *An essay towards a natural history of the corallines...* (privately printed), London, 1755.
156. **Ellis, J. and Solander, D.**, *The natural history of many curious and uncommon zoophytes collected from various parts of the globe*, Benjamin White and Son, London, 1786.
157. **Emery, K. O.**, Marine geology of Johnston Island and its surrounding shallows, Central Pacific Ocean, *Bull. Geol. Soc. Am.*, 67, 1505, 1956.
158. **Farnham, W. F. and Jephson, N. W.**, A survey of the maerl beds of Falmouth (Cornwall), *Brit. Phycol. J.* 12, 119, 1977.
159. **Flajs, G.**, Skeletal structures in some calcifying algae, in *Fossil Algae, Recent Results and Developments*, Flügel, E., Ed., Springer-Verlag, Berlin, 1977, 225.
160. **Foslie, M.**, The Norwegian forms of *Lithothamnion*, *K. Norske Viden. Selsk. Skr.*, 1894, 29, 1895.
160a. **Foslie, M.**, Einiger Bemerkungen über Melobesieae, *Ber. Dtsch. Bot. Ges.*, 15, 252, 1897.
160b. **Foslie, M.**, Weiteres über Melobesieae, *Ber. Dtsch. Bot. Ges.*, 15, 521, 1898.
160c. **Foslie, M.**, Systematical survey of the lithothamnia, *K. Norske Viden. Selsk. Skr.*, 1898, 1898.
160d. **Foslie, M.**, List of species of the lithothamnia, *K. Norske Viden. Selsk. Skr.*, 1898, 1898.
161. **Foslie, M.**, Revised systematical survey of the Melobesieae, *K. Norske Viden. Selsk. Skr.*, 1900, 1900.
161a. **Foslie, M.**, Den botaniska samling, *K. Norske Viden. Selsk. Skr.*, 1902, 23, 1903.
161b. **Foslie, M.**, Lithothamnioneae, Melobesieae, Mastophoreae, in *The Corallinaoeae of the Siboga-Expedition*, Weber-van Bosse, A. and Foslie, M., Siboga-Expeditie, 1904, 61, 10.
162. **Foslie, M.**, New Lithothamnia and systematical remarks, *K. Norske Viden. Selsk. Skr.*, 1905, 1905.
163. **Foslie, M.**, Algologiske notiser V, *K. Norske Viden. Selsk. Skr.*, 1908, 1908.
164. **Foslie, M.**, Algologiske notiser VI, *K. Norske Viden. Selsk. Skr.*, 1909, 1909.
165. **Foslie, M. and Howe, M. A.**, Two new coralline algae from Culebra, Puerto Rico, *Bull. Torrey Bot. Club*, 33, 577, 1906.
166. **Foslie, M.**, *Contributions to a Monograph of the Lithothamnia*, Printz, H., Ed., Trondheim, Norway, 1929.
167. **Foster, M. S.**, Algal succession in a *Macrocystis pyrifera* forest, *Mar. Biol.*, 32, 313, 1975.
168. **Fritsch, F. E.**, *The Structure and Reproduction of the Algae*, Vol. 2., Cambridge University Press, New York, 1945.
169. **Furuya, K.**, Biochemical studies on calcareous algae. I. Major inorganic constituents of some calcareous red algae, *Bot. Mag.*, Tokyo, 73, 355, 1960.
170. **Furuya, K.**, Biochemical studies on calcareous algae. II. Organic acids of some calcareous red algae, *Bot. Mag.*, Tokyo, 78, 274, 1965.
171. **Ganesan, E. K.**, Notes on Indian red algae, II. *Dematolithon ascripticium* (Foslie) Setchell et Mason, *Phykos*, 1, 108, 1962.
172. **Ganesan, E. K.**, Notes on Indian red algae. III. *Fosliella minutula* (Foslie) comb. nov., *Phykos*, 2, 38, 1963.
173. **Ganesan, E. K.**, Studies on the morphology and reproduction of the articulated corallines. I, *Phykos*, 4, 43, 1965.
174. **Ganesan, E. K.**, Morphological studies on the genus *Arthrocardia* Decaisne emend. Areschoug, in Proc. Seminar on Sea, Salt and Plants, Krishnamurthy, V., Eds., Central Salt and Chemicals Research Institute, Bhavnagar, India, 1967, 157.
175. **Ganesan, E. K.**, Studies on the morphology and reproduction of the articulated corallines. III. *Amphiroa* Lamouroux emend. Weber van Bosse, *Phykos*, 6, 7, 1968.
176. **Ganesan, E. K.**, Studies on the morphology and reproduction of the articulated corallines. IV. *Serraticardia* (Yendo) Silva, *Calliarthron* Manza and *Bossiella* Silva, *Bot. Mar.*, 11, 10, 1968.
177. **Ganesan, E. K.**, Studies on the morphology and reproduction of the articulated corallines. II. *Corallina* Linneaus emend. Lamouroux, *Bol. Inst. Oceanogra.*, 7, 65, 1968.
178. **Ganesan, E. K.**, Studies on the morphology and reproduction of the articulated corallines. VI. *Metagoniolithon* Weber van Bosse, *Rev. Algol.*, 10, 248, 1971.
179. **Ganesan, E. K.**, *Amphiroa currae* (Corallinaceae), a new species of marine algae from Venezuela, *Phycologia*, 10, 155, 1971.
180. **Ganesan, E. K., and Desikachary, T. V.**, Studies on the morphology and reproduction of the articulated corallines. V. *Lithothrix* Gray, *Phykos*, 9, 41, 1970.

181. **Garbary, D. J.**, An introduction to the scanning electron microscopy of red algae, in *Modern Approaches to the Taxonomy of Red and Brown Algae*, Systematics Association Special Vol. 10, Irvine, D. E. G. and Price, J. H., Eds., Academic Press, New York, 1978, 205.
182. **Gibbs, R. E.**, *Phyllospadix* as a beach-builder, *Am. Nat.*, 36, 101, 1902.
183. **Gilmartin, M.**, The distribution of the deep water algae of Eniwetok Atoll, *Ecology*, 41, 210, 1960.
184. **Ginsburg, R. N. and Schroeder, J. H.**, Growth and submarine fossilization of algal cup reefs, Bermuda, *Sedimentology*, 20, 575, 1973.
185. **Giraud, G. and Cabioch, J.**, Étude ultrastructurale de l'activité des cellules superficielles du thalle des Corallinacées (Rhodophycées), *Phycologia*, 15, 405, 1976.
186. **Giraud, G. and Cabioch, J.**, Caracteres Généraux de l'ultrastructure des Corallinacées (Rhodophycées), *Rev. Algol.*, 12, 45, 1977.
187. **Gittins, B. T.**, The biology of *Lithothrix aspergillum* J. E. Gray (Corallinaceae, Rhodophyta), Doctoral Dissertation, University of California, Irvine, 1975.
188. **Gittins, B. T. and Dixon, P. S.**, Biology of *Lithothrix aspergillum, Brit. Phyc. J.* 11, 194, 1976.
189. **Gomberg, D. N. and Bonatti, E.**, High-magnesian calcite: leaching of magnesium in the deep sea, *Science*, 168, 1451, 1970.
190. **Gordon, G. C., Masaki, T., and Akioka, H.**, Floristic and distributional account of the common crustose coralline algae on Guam, *Micronesica*, 12, 247, 1976.
190a. **Goreau, T. F.**, On the relation of calcification to primary productivity in reef building organisms, in *Biology of Hydra*, Lenhoff, H. M. and Loomis, W. F., Eds., University of Miami Press, Miami, 1961, 269.
90b. **Goreau, T. F.**, Problems of growth and calcium deposition in reef corals, *Endeavor*, 20, 32, 1961.
191. **Goreau, T. F.**, Calcium carbonate deposition by coralline algae and corals in relation to their roles as reef-builders, *Ann. N. Y. Acad. Sci.*, 109, 127, 1963.
192. **Gray, J. E.**, *Lithothrix*, a new genus of Corallinae, *J. Bot. (London)*, 5, 33, 1867.
193. **Gray, S. F.**, *A Natural Arrangement of British Plants*, Baldwin, Cradock and Joy, London, 1821.
194. **Gross, M. G.**, Carbonate deposits on Plantagenet Bank near Bermuda, *Bull. Geol. Soc. Am.*, 76, 1283, 1965.
195. **Haas, P., Hill, T. G., and Karstens, W. K. H.**, The metabolism of calcareous algae. II. The seasonal variation of metabolic products of *Corallina squamata* Ellis, *Ann. Bot.*, 49, 609, 1935.
196. **Hagerman, L.**, The ostracod fauna of *Corallina officinalis* L. in western Norway, *Sarsia*, 36, 49, 1968.
197. **Hamel, G., and Lemoine, P.**, Corallinacées de France et d'Afrique du Nord, *Arch. Mus. Natl. Hist. Nat. (Paris)*, sér.7, 1, 17, 1953.
198. **Harlin, M. M. and Lindberg, J. M.**, Selection of substrata by seaweeds: optimal surface relief, *Mar. Biol.*, 40, 33, 1977.
199. **Harvey, W. H.**, *Nereis australis...*, Reeve Bros., London, 1847.
200. **Hasagawa, Y. and Wakui, T.**, On the embryological development of the tetraspores in *Corallina pilulifera* P. et R., *Hokkaido Fish. Sci. Inst. Bull.*, 17, 132, 1958.
201. **Hauck, F.**, *Die Meeresalgen*, Edward Kummer, Leipzig, 1885, 575.
201a. **Hawkes, M. W., Tanner, C. E., and Lebednik, P. A.**, The benthic marine algae of northern British Columbia, *Syesis*, 11, 81, 1978.
202. **Hessland, J.**, Uber subfossile Massenworkommen von *Corallina officinalis* Linne, *Senckenbergiana*, 25, 19, 1942.
203. **Heydrich, F.**, Corallinaceae, insbesondere Melobesieae, *Ber. Dtsch. Bot. Ges.*, 15, 34, 1897.
203a. **Heydrich, F.**, Melobesieae, *Ber. Dtsch Bot. Ges.*, 15, 403, 1897b.
204. **Heydrich, F.**, Ueber die weiblichen Conceptakeln von *Sporolithon, Bibl. Bot.*, 49, 1899.
205. **Heydrich, F.**, Weiterer Ausbau des Corallineensystems, *Ber. Dtsch. Bot. Ges.*, 18, 310, 1900.
206. **Heydrich, F.**, Carpogonium und Auxiliarzelle einiger Melobesieae, *Ber. Dtsch Bot. Ges.*, 27, 79, 1909.
207. **Hicks, G. R. F.**, Checklist and ecological notes on the fauna associated with some littoral corallinacean algae, *Bull. Nat. Sci. Victoria Univ., Wellington Biol. Soc., Wellington, N.Z.*, 2, 47, 1971.
208. **Hollenberg, G. J.**, Phycological notes IV, including new marine algae and new records for California, *Phycologia*, 9, 61, 1970.
209. **Hommersand, M. H.**, Taxonomic and phytogeographic relationships of warm temperate marine algae occurring in Pacific North America and Japan, in *Proc. 7th Int. Seaweed Symp.* Nizizawa, K., Ed., University of Tokyo Press, Tokyo, 1972, 66.
210. **Howe, M. A.**, Some of the coralline seaweeds in the museum, *J. N. Y. Bot. Gard.*, 6, 59, 1905.
211. **Howe, M. A.**, The building of "coral" reefs, *Science*, 22, 837, 1912.
212. **Howe, M. A.**, On some fossil and recent Lithothamnieae of the Panama Canal Zone, *U. S. Nat. Mus. Bull.*, 103, 1, 1919.

213. **Howe, M. A., Algae,** in *The Bahama Flora,* Britton, N. L. and Millspaugh C. F., Eds., Hafner, New York, 1920, 553.
214. **Huvé, H.,** Contribution a l'étude des fonds a *Lithothamnium (?) solutum* Foslie (= *Lithophyllum solutum* (Foslie) Lemoine) de la region de Marseille, *Rec. Trav. Stn. Mar. Endoume - Marseille Fasc. Hors Ser. Suppl.,* 18, 105, 1956.
215. **Huvé, H.,** Sur l'induvidualité générique du *Tenarea undulosa* Bory 1832 et du *Tenarea tortuosa* (Esper) Lemoine 1911, *Bull. Soc. Bot. Fr.,* 104, 132, 1957.
216. **Huvé, H.,** Taxonomie, écologie et distribution d'une Melobesiée méditérranéenne: *Lithophyllum papillosum* (Zanardini) comb. nov., non *Lithophyllum (Dermatolithon) papillosum* (Zanard.) Foslie, *Bot. Mar.,* 4, 219, 1962.
217. **Huvé, P.,** Sur la reinstallation d'un "Trottoir à *Tenarea,*" en Méditérranée occidentale, *C. R. Acad. Sci., Ser. D,* 253, 2157, 1956.
218. **Huvé, P.,** Resultats preliminaires de l'étude experimentale de la reinstallation de peuplements a *Lithophyllum incrustans* Ph., Repp P. - V. Reun. C.I.E.S.M.M.,15, 65, 1960.
219. **Ishijima, W.,** Tertiary and Pleistocene algae from Mindoro, the Phillippines, *Geol. Palaeontol. Southeast Asia,* 6, 277, 1969.
220. **Jacquotte, R.,** Étude des fonds des maerl de Méditérranée, *Rec. Trav. Sta. Mar. D'Endoume, Bull.,* 26, 141, 1962.
221. **Johansen, H. W.,** A new member of the Corallinaceae: *Chiharaea bodegensis* gen. et sp. nov., *Phycologia,* 6, 51, 1966.
222. **Johansen, H. W.,** Reproduction of the articulated coralline *Amphiroa ephedraea, J. Phycol.,* 4, 319, 1968.
223. **Johansen, H. W.,** Morphology and systematics of coralline algae with special reference to *Calliarthron, Univ. Calif. Berkeley, Publ. Bot.,* 49, 1969.
224. **Johansen, H. W.,** Patterns of genicular development in *Amphiroa* (Corallinaceae), *J. Phycol.,* 5, 118, 1969.
225. **Johansen, H. W.,** The diagnostic value of reproductive organs in some genera of articulated coralline algae, *Br. Phycol. J.,* 5, 79, 1970.
226. **Johansen H. W.,** Changes and additions to the articulated coralline flora of California, *Phycologia,* 10, 241, 1971.
227. **Johansen, H. W.,** *Bossiella,* a genus of articulated corallines (Rhodophyceae, Cryptonemiales) in the eastern Pacific, *Phycologia,* 10, 381, 1971.
228. **Johansen, H. W.,** Effects of elevation changes on benthic algae in Prince William Sound, in *The Great Alaska Earthquake of 1964: Biology,* National Academy of Sciences, Washington, D.C., 1971, 35.
229. **Johansen, H. W.,** Conceptacles in the Corallinaceae, in *Proc. 7th Int. Seaweed Symp.* Nizizawa, K., Ed., University of Tokyo Press, Tokyo, 1972, 114.
230. **Johansen, H. W.,** Ontogeny of sexual conceptacles in a species of *Bossiella* (Corallinaceae), *J. Phycol.,* 9, 141, 1973.
231. **Johansen, H. W.,** Articulated coralline algae, *Oceanogr. Mar. Biol.,* 12, 77, 1974.
232. **Johnsen, H. W.,** Phycological Reviews 4. Current status of generic concepts in coralline algae (Rhodophyta), *Phycologia,* 15, 221, 1976.
233. **Johansen, H. W.,** The articulated Corallinaceae (Rhodophyta) of South Africa. I. *Cheilosporum* (Decaisne) Zanardini, *J. S. Afr. Bot.,* 43, 163, 1917.
234. **Johansen, H. W. and Austin, L. F.,** Growth rates in the articulated coralline *Calliarthron* (Rhodophyta), *Can. J. Bot.,* 48, 125, 1970.
235. **Johansen, H. W. and Colthart, B. J.,** Variability in articulated coralline algae (Rhodophyta), *Nova Hedwigia Z. Kryptogamenkd,* 26, 135, 1975.
235a. **Johansen, H. W., Irvine, L. M., and Webster, A. M.,** *Haliptylon squamatum* (L.) comb. nov., a poorly known British coralline alga, *Br. Phycol. J.,* 8, 212, 1973.
236. **Johansen, H. W. and Silva, P. C.,** Janieae and Lithotricheae: two new tribes of articulated Corallinaceae (Rhodophyta), *Phycologia,* 17, 413, 1978.
237. **John, D. M. and Lawson, G. W.,** Observations on the marine algal ecology of Gabon, *Bot. Mar.,* 17, 249, 1974.
238. **Johnson, J. H.,** Ancestry of the coralline algae, *J. Paleontol.* 30, 563, 1956.
239. **Johnson, J. H.,** The algal genus *Archaeolithothamnium* and its fossil representatives, *J. Paleontol.,* 37, 175, 1963.
240. **Johnson, J. H.,** *Limestone-building Algae and Algal Limestones,* Johnson Publishing Co., Boulder, 1961, 297.
241. **Jones, W. E. and Moorjani, S. A.,** The attachment and early development of the tetraspores of some coralline red algae, *Spec. Publ. Mar. Biol. Assc. India,* 1973, 293, 1973.

242. **Joly, A. B.**, Flora marinha do litoral norte do Estado de São Paulo e regiões circunvizinhas, *Fac. Filos. Ciênc. Let., Univ. Sao Paulo, Bol. Bot.*, 21, 5, 1965.
243. **Joly, A. B., Cordeiro, M., Yamaguishi, N. and Ugadim, Y.**, New marine algae from southern Brazil, *Rickia*, 2, 159, 1965.
244. **Katada, M. and Matsui, T.**, Studies on the vertical distribution and succession of intertidal algal communities. I, *Bull. Soc. Plant Ecol.*, 3, 17, 1953.
245. **Katada, M. and Matsui, T.**, Studies on the repopulation of the tidal vegetation. II, *Bull. Soc. Plant Ecol.*, 3, 153, 1954.
246. **Keith, M. L. and Weber, J. N.**, Systematic relationships between carbon and oxygen isotopes in carbonates deposited by modern corals and algae, *Science*, 150, 498, 1965.
247. **Kindig, A. C.**, Physiological responses of sewage-resistant macrophytes to effluent stress, *Influence of Domestic Wastes on the Structure and Energenetics of Intertidal Communities near Wilson Cove, San Clemente Island,* Littler, M. M. and Murray, S. N., Eds., California Water Resources Center, University of California, Davis, 164, 1977, 69.
248. **Kitching, J. A.**, Studies in sublittoral ecology. II. Recolonization at the upper margin of the sublittoral region; with a note on the denudation of a *Laminaria* forest by storms, *J. Ecol.*, 25, 482, 1937.
249. **Kohlmeyer, J.**, Parasitische and epipytische Pilze auf Meersalgen, *Nova Hedwigia Z. Kryptogamenkd.*, 5, 127, 1963.
250. **Kohlmeyer, J.**, Intertidal and phycophilous fungi from Tenerife (Canary Islands), *Trans. Br. Mycol. Soc.*, 50, 137, 1967.
250a. **Kützing, F. T.**, Phycologia generalis; oder Anatomie, Physiologie und Systemkunde der Tange, *Leipzig, 1843.*
250b. **Kützing, F. T.**, *Species algarum*, Brockhaus, F. A., Leipzig, 1849.
251. **Kützing, F. T.**, *Tabulae phycologicae oder Abbildungen der Tange*, Vol. 8, Nordhausen, 1858.
252. **Kylin, H.**, Entwicklungsgeschichtliche Florideenstudien, *Acta Univ. Lund., Sect. 2*, 24, 1, 1928.
253. **Kylin, H.**, *Die gattungen der Rhodophyceen*, Gleerups, Lund, 1956.
254. **Lamarck, J. B.**, Sur les poplyiers corticifères, *Mem. Mus. Hist. Nat. Paris*, 2, 76, 1815.
255. **Lamouroux, J. V. F.**, Extrait d'un mémoire sur la classification des Polypiers coralligènes, non entièrement pierreux..., *Nouveau Bull. Sci. Soc. Philomatique Paris*, 3, 181, 1812.
256. **Lamouroux, J. V. F.**, *Histoire des polypiers coralligènes flexibles, vulgairement nommés zoophytes*, (Publ. by author), Caen, 1816.
257. **Lamouroux, J. V. F.**, Description des Polypiers flexibles, in *Voyage Autour du Monde*, de Freycinet, L., Ed., Paris, 1824, 621.
258. **Lancelot, A.**, Les algues Mélobésiées de Biarritz, *Bull. Cent. Etud. Rech. Sci., Biarritz*, 3, 505, 1961.
259. **Lawrence, J. M. and Dawes, C. J.**, Algal growth over the epidermis of sea urchin spines, *J. Phycol.*, 5, 269, 1969.
260. **Lebednik, P. A.**, Ecological effects of intertidal uplifting from nuclear testing, *Mar. Biol.*, 20, 197, 1973.
261. **Lebednik, P. A.**, A new record of *Chiharaea* Johansen (Rhodophycophyta, Corallinaceae) from British Columbia, *Syesis*, 8, 397, 1976.
262. **Lebednik, P. A.**, The Corallinaceae of northwestern North America. I. *Clathromorphum* Foslie emend. Adey, *Syesis*, 9, 59, 1977.
263. **Lebednik, P. A.**, Postfertilization development in *Clathromorphum, Melobesia* and *Mesophyllum* with comments on the evolution of the Corallinaceae and the Cryptonemiales (Rhodophyta), *Phycologia*, 16, 379, 1977.
264. **Lebednik, P. A.**, Addendum to Postfertilization development in *Clathromorphum, Melobesia* and *Mesophyllum* with comments on the evolution of the Corallinaceae and the Cryptonemiales (Rhodophyta), *Phycologia*, 17, 358, 1978.
265. **Lebednik, P. A.**, Development of male conceptacles in *Mesophyllum* Lemoine and other genera of the Corallinaceae (Rhodophyta), *Phycologia*, 17, 388, 1978.
266. **Lee, R. K. S.**, Taxonomy and distribution of the melobesioid algae on Rongelap Atoll, Marshall Islands, *Can. J. Bot.*, 45, 985, 1967.
267. **Lee, R. K. S.**, Developmental morphology of the crustaceous alga *Melobesia mediocris*, *Can. J. Bot.*, 48, 437, 1970.
268. **Leighton, D. L. and Boolootian, R. A.**, Diet and growth in the black abalone *Haliotis cracherodii*, *Ecology*, 44, 227, 1963.
269. **Lemoine, M.**, Structure anatomique des Mélobésiées. Application à la classification, *Ann. Inst. Océanogr. Monaco*, 2, 1, 1911.
270. **Lemoine, M.**, Un nouveau genre de Mélobésiées: *Mesophyllum*, *Bull. Soc. Bot. Fr.* 75, 251, 1928.
271. **Lemoine, M.**, Les corallinacées de l'Archipel des Galapagos et du Golfe de Panama, *Arch. Mus. Hist. Nat.*, 4, 27, 1929.
272. **Lemoine M.**, Les algues calcaires de la zone neritique, *Soc. Biogr.*, 7, 75, 1940.

273. **Lemoine, M.**, Apparition de la structure monostromatique dans un thalle epais de *Dermatolithon* (Mélobésiées, Corallinacées), *Bull. Soc. Bot. Fr.*, 117, 547, 1970.
274. **Lemoine, M.**, Sur un processus d'edophytisme auto-spécifique dans une algue calaire (*Lithoporella*, Mélobésiées, Corallinacées), *C. R. Acad. Sci., Ser. D*, 270D, 2645, 1970.
275. **Lemoine, M.**, Contribution a l'étude du genre *Lithoporella* (Corallinacées), *Rev. Algol.*, 11, 42, 1974.
276. **Lemoine, M.**, Les difficultés de la phylogénie chez le Algues Corallinacées, *Bull. Soc. Geol. Fr.*, 19, 1319, 1977.
277. **Lemoine, M., and Emberger, J.**, *Kymalithon*, nouveau genre de Mélobésiée de L'Aptien supérieur et considérations sur l'âge du faciès a Mélobésiées dit "faciès ce Vimport," *Actes Soc. Linn. Bordeaux, Ser. B*, 104, 14, 1967.
278. **Lewalle, J.**, Détermination macroscopique des Algues rouges calcaires (Corallinaceae et Squamariaceae partim) du golfe de Naples, *Pubbl. Stn. Zool. Napoli*, 32, 241, 1961.
279. **Lewis, J. B**, Processes of organic production on coral reefs, *Biol. Rev.*, 52, 305, 1977.
280. **Linnaeus, C.**, *Systema Naturae...*, Vol. 1, 10th ed., L. Salvii, Stockholm, 1758.
281. **Linnaeus, C.**, *Systema Naturae*, Vol. 1, 12th ed., L. Salvii, Stockholm, 1767.
282. **Littler, M. M.**, Standing stock measurements of crustose coralline algae (Rhodophyta) and other saxicolous organisms., *J. Exp. Mar. Biol. Ecol.*, 6, 91, 1971.
283. **Littler, M. M.**, *Tenarea tessalatum* (Lemoine) Littler, comb. nov., an unusual crustose coralline (Rhodophyceae, Crytpoemiales) from Hawaii, *Phycologia*, 10, 355, 1971.
284. **Littler, M. M.**, The crustose Corallinaceae, *Oceanog. Mar. Biol.*, 10, 311, 1972.
285. **Littler, M. M.**, The population and community structure of Hawaiian fringing-reef crustose Corallinaceae (Rhodophyta, Cryptonemiales), *J. Exp. Mar. Biol. Ecol.*, 11, 103, 1973.
286. **Littler, M. M.**, The distribution, abundance, and communities of deepwater Hawaiian crustose Corallinaceae (Rhodophyta, Cryptonemiales), *Pac. Sci.*, 27, 281, 1973.
287. **Littler, M. M.**, The productivity of Hawaiian fringing-reef crustose Corallinaceae and an experimental evaluation of production methodology, *Limnol. Oceanogr.*, 18, 946, 1973.
288. **Littler, M. M.**, Calcification and its role among the macroalgae, *Micronescia J. Coll. Guam*, 12, 27, 1976.
289. **Littler, M. M. and Doty, M. S.**, Ecological components structuring the seaweed edges of tropical Pacific reefs: the distribution, communities and productivity of *Porolithon*, *J. Ecol.*, 63, 117, 1975.
290. **Littler, M. M. and Murray, S. N.**, The primary productivity of marine macrophytes from a rocky intertidal community, *Mar. Biol.*, 27, 131, 1974.
291. **Littler, M. M. and Murray, S. N.**, Impact of sewage on the distribution, abundance and community structure of rocky intertidal macro-organisms. *Mar. Biol.*, 30, 277, 1975.
291a. **Littler, M. M., Murray, S. N., and Arnold, K. E.**, Seasonal variations in net photosynthetic performance and cover of intertidal macrophytes, *Aquatic Bot.*, 7, 35, 1979.
292. **Lund, S.**, The marine algae of East Greenland. I. Taxonomical part, *Medd. Groenl.*, 156, 247, 1959.
293. **Magne, F.**, Recherches caryologiques chez les Floridées (Rhodophycées), *Cah. Biol. Mar.*, 5, 461, 1964.
294. **Manza, A. V.**, The genera of articulated corallines, *Proc. Natl. Acad. Sci. U.S.A.*, 23, 44, 1937.
295. **Manza, A. V.**, Some North Pacific species of articulated corallines, *Proc. Natl. Acad. Sci. U.S.A.*, 23, 561, 1937.
300. **Manza, A. V.**, New species of articulated corallines from South Africa, *Proc. Natl. Acad. Sci. U.S.A.*, 23, 568, 1937.
301. **Manza, A. V.**, A revision of the genera of articulated corallines, *Philipp. J. Sci.*, 71, 239, 1940.
302. **Marsh, J. A.**, Primary productivity of reef-building calcareous red algae, *Ecology*, 51, 255, 1970.
303. **Masaki, R.**, Studies on the Melobesioideae of Japan, *Mem. Fac. Fish., Hokkaido Univ.*, 16, 1968.
304. **Masaki, T. and Tokida, J.**, Studies on the Melobesioideae of Japan. II, *Bull. Fac. Fish., Hokkaido Univ.*, 10, 285, 1960.
305. **Masaki, T. and Tokida, J.**, Studies on the Melobesioideae of Japan. III, *Bull. Fac. Fish., Hokkaido Univ.*, 11, 37, 1960.
306. **Masaki, R. and Tokida, J.**, Studies on the Melobesioideae of Japan. IV, *Bull. Fac. Fish., Hokkaido Univ.*, 11, 188, 1961.
307. **Masaki, T. and Tokida, J.**, Studies on the Melobesioideae of Japan. V, *Bull. Fac. Fish., Hokkaido Univ.*, 12, 161, 1961.
308. **Masaki, T. and Tokida, J.**, Studies on the Melobesioideae of Japan. VI, *Bull. Fac. Fish., Hokkaido Univ.*, 14, 1, 1963.
309. **Mason, L. R.**, The crustaceous coralline algae of the Pacific Coast of the United States, Canada and Alaska, *Univ. Calif., Berkeley, Publ. Bot.*, 26, 313, 1953.
310. **Massieux, M.**, A comparison of the anatomical structures of a recent and a fossil species of the Corallinaceae, in *Fossil Algae, Recent Results and Developments*, Flügel, E., Ed., Springer-Verlag, Berlin, 1977, 190.

311. **Mastrorilli, V. I.**, Caratteri morfologici e structturali di un esemplare fertile de *"subterraniphyllum"* Elliott Rinvenuto nell'Oligocene di ponzone (Piemonte), *Riv. Ital. Paleont. Stratigr.*, 74, 1275, 1968.
312. **McCandless, E. L.**, The importance of cell wall consituents in algal taxonomy, in *Modern Approaches to the Taxonomy of Red and Brown Algae*, Systematics Association Special Vol. 10, Irvine, D. E. G. and Price, J. H. Eds., Academic Press, New York, 1978, 63.
313. **McLean, J. H.**, Sublittoral ecology of kelp beds of the open coast area near Carmel, California, *Biol. Bull.*, 122, 95, 1962.
314. **McMaster, R. L. and Conover, J. T.**, Recent algal stromatolites from the Canary Islands, *J. Geol.*, 74, 647, 1966.
315. **Meischner, D. and Meischner, U.**, Bermuda south shore reef morphology. A preliminary report, in *Proc. 3rd Int. Coral Reef Symp.*, Rosenstiel School of Marine and Atmospheric Science, University of Miami, Miami, 1977, 244.
316. **Mendoza, M. L.**, Distribution de quelques espèces de Corallinacées articulées sur les côtes d'Argentine, *Bull. Soc. Phycol. Fr.*, 19, 67, 1974.
317. **Mendoza, M. L.**, Estudio de las variacones morfologicas externas, internas y citologicas de la Corallineae (Rhodophyta) de la Argentina, *Physis*, 35, 15, 1976.
318. **Mendoza, M. L.**, *Antarcticophyllum*, nuevo genero para las Corallinaceae, *Bol. Soc. Argent. Bot.*, 17, 252, 1976.
319. **Meslin, R. E.**, Sur la position et la valeur taxonomique de *Corallina elegans* Lenormand (Rhodophyceés, Cryptonemiales), *Phycologia*, 15, 415, 1976.
320. **Milliman, J. D.**, *Marine carbonates*, Springer-Verlag, Berlin, 1974.
321. **Milliman, J. D.**, Role of calcareous algae in Atlantic continental margin sedimentation, in *Fossil Algae, Recent Results and Developments*, Flügel, E., Ed., Springer-Verlag, Berlin, 1977, 232.
322. **Milliman, J. . and Emery, K. O.**, Sea levels during the past 35,000 years, *Science*, 162, 1121, 1968.
323. **Milliman, J. D., Gastner, M., and Muller, J.**, Utilization of magnesium in coralline algae, *Geol. Soc. Am. Bull.*, 82, 573, 1971.
324. **Minder, F.**, Die Fruchtentwicklung von *Choreonema thureti*, Dissertation, Freiberg, 1910.
325. **Munda, I.**, On the chemical composition, distribution and ecology of some common benthic marine algae from Iceland, *Bot. Mar.*, 15, 1, 1972.
326. **Murata, K. and Masaki, T.**, Studies of reproductive organs in articulated coralline algae of Japan, *Phycologia*, 17, 403, 1978.
327. **Murray, S. N. and Littler, M. M.**, Seasonal analyses of standing stock and community structure of mca macro-organisms, in *Influence of Domestic Wastes on the Structure and Energetics of Intertidal Communities Near Wilson Cove, San Clemente Island*, California Water Resources Center, University of California, Davis, 164, 1977.
327a. **Myers, A. and Preston, R. D.**, Fine structure in the red algae III. A general survey of cell-wall structure in the red algae, *Proc. R. Soc. London, Ser. D*, 150B, 456, 1959.
328. **Naylor, G. L. and Russell-Wells, B.**, On the presence of cellulose and its distribution in the cell-walls of brown and red algae, *Ann. Bot. (London)*, 48, 635, 1934.
329. **Nelson, R. J. and Duncan, Prof.**, On some points in the histology of certain species of Corallinaceae, *Trans. Linn. Soc. London Bot., Ser. 2*, 1, 197, 1876.
330. **Neushul, M.**, Functional interpretation of benthic marine algae morphology, in *Contributions to the Systematics of Benthic Marine Algae of the North Pacific*, Abbott, I. , and Kurogi, M., Eds., 1972, 47.
330a. **Ngan, Y. and Price, I. R.**, Systematic significance of spore size in the Florideophyceae (Rhodophyta), *Br. Phycol. J.*, 14, 285, 1979.
331. **Notoya, M.**, Spore germination in crustose coralline *Tenarea corallinae, T. dipar* and *T. tumidula*, *Bull. Jpn. Soc. Phycol.*, 22, 47, 1974.
332. **Notoya, M.**, Spore germination in several species of crustose corallines (Corallinaceae, Rhodophyta), *Bull. Fac. Fish., Hokkaido Univ.*, 26, 314, 1976.
333. **Notoya, M.**, On the influence of various culture conditions on the early development of spore germination in three species of the crustose corallines (Rhodophyta) (Preliminary report), *Bull. Jpn. Soc. Phycol.*, 24, 137, 1976.
334. **Odum, H. T. and Odum, E. P.**, Trophic structure and productivity of a windward coral reef community on Eniwetok Atoll, *Ecol. Monogr.*, 24, 291, 1955.
335. **Ogata, E.**, On the distribution of spores from the community of coralline algae, *Jap. J. Ecol., (Nippon Seitai Gakkaishi)*, 2, 104, 1952.
336. **Ogata, E.**, On the distribution of spores from the community of coralline algae-supplementary report, *Bull. Soc. Plant Ecol.*, 3, 60, 1953.
337. **Ogata, E.**, Some experiments on the settling of spores of red algae, *Jpn. J. Ecol.*, 3, 128, 1953.
338. **Okamura, K.**, *Nipon Kaisoshi, (Marine Algae of Japan)*, Uchidaro-Kakuho, Tokyo, 1936.
339. **Okazaki, M.**, Carbonic anhydrase in the calcareous red alga, *Serraticardia maxima, Bot. Mar.*, 15, 133, 1972.

340. Okazaki, M., Some enzymatic properties of $^{2+}$-dependent adenosine triphosphatase from a calcareous red algae, *Serraticardia maxima*, and its distribution in marine algae, *Bot. Mar.*, 20, 347, 1977.
341. Okazaki, M. and Furuya, K., Carbonic anhydrase in algae, in *Proc. 7th Int. Seaweed Symposium*, 1971, Nizizawa, K., Ed., University of Tokyo Press, Tokyo, 1972, 522.
342. Okazaki, M., Ikawa, T., Furuya, K., Nisizawa, K., and Miwa, T., Studies on calcium carbonate deposition of a calcareous red alga *Serraticardia maxima*, *Bot. Mag.*, Tokyo, 83, 193, 1970.
343. Orszag-Sperber, F., Poignant, F., and Poisson, A., Paleogeographic significance of rhodolites: some examples from the Miocene of France and Turkey, in *Fossil Algae, Recent Results and Developments*, Flügel, E., Ed., Springer-Verlag, Berlin, 1977, 286.
344. Paine, R. T. and Vadas, R. L., The effects of grazing by sea urchins, *Strongylocentrotus* spp., on benthic algal populations, *Limnol. Oceanogr.*, 14, 710, 1969.
344a. Pallas, P. S., Elenchus zoophytorum, *P. van Cleef*, The Hague, 1766.
345. Papenfuss, G. F., Notes of South African marine algae III *J. S. Afr. Bot.*, 17, 167, 1952.
346. Papenfuss, G. F., A review of the present system of classification of the Florideophycidae, *Phycologia*, 5, 247, 1966.
347. Papenfuss, G. F., A history catalogue, and bibliography of Red Sea benthic algae, *Isr. J. Bot.* 17, 1968.
348. Parke, M. and Dixon, P. S., Check-list of British Marine algae, *J. Mar. Biol. Assoc., U. K.*, 56, 527, 1976.
349. Pearse, V., Radioisotopic study of calcification in the articulated coralline alga *Bossiella orbigniana*, *J. Phycol.*, 8, 88, 1972.
350. Peel, M. C. and Duckett, J. G., Studies of spermatogenesis in the Rhodophyta, *Biol. J. Linn. Soc.*, 7, 1, 1975.
351. Peel, M. C., Lucas, I. A. N., Duckett, J. G., and Greenwood, A. D., Studies of sporogenesis in the Rhodophyta. I. An association of the nuclei with endoplasmic reticulum in post-meiotic tetraspore mother cells of *Corallina officinalis* L., *Z. Zellforsch. Microsk. Anat.*, 147, 59, 1973.
352. Pequegnat, W. E., Population dynamics in a sublittoral epifauna, *Pac. Sci.*, 17, 424, 1963.
353. Philippi, R. A., Beweis, dass die Nulliporen Pflanzen sind. *ch. Naturgesch*, 3, 387, 1837.
354. Pilger, R., Ein Beitrag zur Kenntnis der rallinaceae, *Bot. Jahrb.*, 41, 241, 1908.
356. Poignant, F., Une Nouvelle Algue Corallinacee, *Neogoniolithon montainvillense* n. sp. Dans Le Paleocene du Bassin Parisien, *Géobios*, 10, 129, 1977.
357. Poignant, A. F., Les Algues fossiles, point de vue du geologue, *Soc. Phycol. Fr., Bull.*, 22, 87, 1977.
357a. Poignant, A. F., The Mesozoic red algae: a general survey, in *Fossil Algae. Recent Results and Developments*, Flügel, E., Ed., Springer-Verlag, Berlin, 1977, 177.
358. Poisson, A. and Poignant, F., La formation de Karabayir, base de la transgression miocene dans la region de Korkuteli epartment d'talayaTurquie). *Lithothamnium pseudoamossissimum*, nouvelle espece d'algue rouge de la formation de Karabayir, *Bull. Min. Expl. Inst. Turkey*, 82, 67, 1974.
359. Pollock, J. B., Fringing and fossil coral reefs of Oahu, Bull. *Bernice P. Bishop Mus. Spec. Publ.* 55, 1928.
360. Postels, A. and Ruprecht, F., *Illustrationes algarum*...Leningrad, 1840.
361. Pujals, C., Catalogo de Rhodophyta citadas para la Argentina, Rev. Mus. Argent. Cienc. Nat. "Bernardino Rivadavia" *Inst. Nat. Invest. Cienc. Nat., Cienc. Bot.*, 3, 3, 1963.
362. Pujals, C., Adiciones y correcciones al "Catalogo de Rhodophyta citadas para la Argentina", Rev. Mus. Argent. Cienc. Nat. "Bernardino Riviadavia" *Inst. Nac. Invest. Cienc. Nat., Cienc. Bot.*, 5, 123, 1977.
363. Riding, R., Problems of affinity in Palaezoic calcareous algae, in *Fossil Algae, Recent Results and Development*, Flügel, E., Ed., Springer-Verlag, Berlin, 1977, 202.
364. Rosanoff, S., Recherches anatomiques sur les Mélobésiées, *Mem. Soc, Sci. Nat. Cherbourg*, 12, 1, 1866.
365. Rosenvinge, L. K., The marine algae of Denmark.Contributions to their natural history., II. Rhodophyceae II. (Cryptomeniales)., *K. Danske Vid. Selsk. Skr., Naturvidensk. Math. Afd.*, 7, 153, 1917.
365a. Ross, A. G., Some typical analyses of red seaweeds, *J. Sci. Food Agric.*, 4, 333, 1953.
365b. Rothpletz, A., Fossile kalkalgen aus den Familien der diaceen und der rallineen, *Z. Dtsch Geol. Ges.*, 43, 295, 1891.
366. Rothpletz, A., Ueber eine neue Pflanze (*Lithothamnium erthraeum* n. sp.) des Rothen Meeres, *Bot. Centralbl.*, 54, 5, 1893.
367. Saito, Y., Sasaki, H., and Watanabe, K., Succession of algal communities on the vertical substratum faces of breakwaters in Japan, *Phycologia*, 15, 93, 1976.
368. Santelices, B. and Abbott, I. A., New records of marine algae from Chile and their effect on phytogeography, *Phycologia*, 17, 213, 1978.
369. Schmalz, R. F., Brucite in carbonate secreted by the red alga *Goniolithon* sp., *Science*, 149, 993, 1965.

370. **Schmitz, F.**, Systematische übersicht der bisher bekannten Gattungen der Florideen, *Flora*, 72, 435, 1889.
371. **Schmitz, F., and Hauptfleisch, P., Corallinaceae,** in *Die Naturlichen Pflanzenfamilien ...Teil I, Abt.* 2, Engler, A. and Prantl, K., Eds., Leipzig, 1897, 537.
372. **Schwab, K. W.**, Calcareous red algae of the vicinity of Puerto Penasco, Sonora, Mexico. Morphology of *Lithophyllum pallescens* (Foslie) Heydrich, *J. Ariz. Acad. Sci.*, 5, 189 1969.
373. **Seagreif, S. C.**, *The seaweeds of the Tsitsikama Coastal National Park,* National Parks Board, Republic of South Africa, 1967.
373a. **Seapy, R. R. and Littler, M. M.**, The distribution, abundance, community structure, and primary productivity of macroorganisms from two central California rocky intertidal habitats, *Pac. Sci.*, 32, 293, 1978.
374. **Searles, R. B. and Schneider, C. W.**, A checklist and bibliography of North Carolina seaweeds, *Bot. Mar.*, 21, 99, 1978.
375. **Sears, J. R. and Cooper, R.**, Descriptive ecology of offshore, deepwater, benthic algae in the temperate western North Atlantic Ocean, *Mar. Biol.*, 44, 309, 1978.
376. **Sears, J. R. and Wilce, R. T.**, Sublittoral, benthic marine algae of Southern Cape Cod and adjacent islands: seasonal periodicity, associations, diversity, and floristic composition, *Ecol. Monogr.*, 45, 337, 1975.
377. **Segawa, S.**, Systematic anatomy of the articulated corallines. I. *Amphiroa rigida* Lamouroux, *J. Jpn. Bot.*, 16, 219, 1940.
378. **Segawa, S.**, Systematic anatomy of the articulated corallines. II. *Amphiroa misakiensis* Yendo., *J. Jpn. Bot.*, 16, 488, 1940.
379. **Segawa, S.**, Systematic anatomy of the articulated corallines. III. *Amphiroa aberrans* Yendo, *J. Jpn. Bot.*, 17, 164, 1941.
380. **Segawa, S.**, Systematic anatomy of the articulated corallines. IV. *Amphiroa crassissima* Yendo., *J. Jpn. Bot.*, 17, 226, 1941.
381. **Segawa, S.**, Systematic anatomy of the articulated corallines. V. *Amphiroa cretacea* (Postels et Ruprecht) Endlicher, *J. Jpn. Bot.*, 17, 348, 1941.
382. **Segawa, S.**, Systematic anatomy of the articulated corallines. VI. *Cheilosporum jungermannioides* Ruprecht) Areschoug, *J. Jpn. Bot.*, 17, 450, 1941.
383. **Segawa, S.**, Systematic anatomy of the articulated corallines. VII. *Cheilosporum yessoense* Yendo, *J. Jpn. Bot.*, 17, 463, 1941.
384. **Segawa, S.**, Systematic anatomy of the articulated corallines. VIII. *Cheilosporum maximum* Yendo, *J. Jpn. Bot.*, 17, 647, 1941.
385. **Segawa, S.**, Systematic anatomy of the articulated corallines. IX. *Corallina* sp., *J. Jpn. Bot.*, 18, 573, 1942.
386. **Segawa, S.**, Systematic anatomy of the articulated corallines. X. *Jania radiata* Yendo, *Seibutsu*, 1, 151, 1946.
387. **Segawa, S.**, Systematic anatomy of the articulated corallines. XI. *Lithothrix aspergillum* Gray, *Seibutsu*, 2, 87, 1947.
388. **Segawa, S.**, Systematic anatomy of the articulated corallines. XII. *Metagoniolithon, Arthrocardia* and *Duthiea, Seibutsu,* 4, 52, 1949.
389. **Segawa, S.**, Systematic anatomy of the articulated corallines. The structure and reproduction of *Yamadaia melobesioides* Segawa, *Bot. Mag.*, Tokyo, 68, 241, 1955.
390. **Segawa, S.**, Studies on the corallinaceous algae in the warmer seas around Japan, *Records Oceanogr. Works Jpn.*, 3, 221, 1959.
391. **Setchell, W. A.**, The coral reef problem in the Pacific, *Proc. 3rd Pan-Pacific Sci. Congr., Tokyo*, 323, 1926.
392. **Setchell, W. A.**, Botanical view of coral reefs especially of those of the Indo-Pacific region, *Proc. 3rd Pan-Pacific Sci. Congr., Tokyo*, 1837, 1926.
393. **Setchell, W. A.**, Nullipore versus coral in reef-formation, *Proc. Am. Philos. Soc.*, 65, 136, 1926.
394. **Setchell, W. A.**, Nullipore reef control and its significance, *Prog. 4th Pac. Sci. Cgr.*, Java, 265, 1929.
395. **Setchell, W. A.**, Biotic cementation in coral reefs, *Proc. Natl. Acad. Sci. U.S.A.*, 16, 781, 1930.
396. **Setchell, W. A.**, *Mastophora* and the Mastophoreae: genus and subfamily of Corallinaceae, *Proc. Natl. Acad. Sci. U.S.A.*, 29, 127, 1943.
397. **Setchell, W. A., and Gardner, N. L.**, Algae of northwestern America, *Univ. Calif. Publ. Bot.*, 1, 165, 1903.
398. **Setchell, W. A., and Gardner, N. L.**, The marine algae of the Revillagigedo Islands expedition in 1925, *Proc. Calif. Acad. Sci., Ser. 4,* 19, 109, 1930.
399. **Setchell, W. A. and Mason, L.**, New or little known crustacious corallines of Pacific North America, *Proc. Natl. Acad. Sci. U.S.A.*, 29, 92, 1943.

400. Setchell, W. A. and Mason, L., *Goniolithon* and *Neogoniolithon:* two genera of crustaceous coralline algae, *Proc. Natl. Acad. Sci. U.S.A.*, 29, 87, 1943.
401. Shepherd, S. and Womersley, H. B. S., The sublittoral ecology of West Island, South Australia. I. Environmental features and algal ecology, *Trans. R. Soc. South Aust.*, 94, 105, 1970.
402. Shepherd, S. A. and Womersley, H. B. S., Pearson Island Expedition 1969. The sub-tidal ecology of benthic algae, *Trans. R. Soc. South Aust.*, 95, 155, 1971.
403. Shepherd, S. and Womersley, H. B. S., The subtidal algal and seagrass ecology of St. Francis Island, South Australia, *Trans. R. Soc. South Aust.*, 100, 177, 1976.
404. Siesser, W. G., Relief algal nodules (rhodolites) from the South African continental shelf, *J. Geol.*, 80, 611, 1972.
405. Silva, P. C., Notes on Pacific marine algae, *Madroño*, 14, 41, 1957.
406. Silva, P. C., Status of our knowledge of the Galapagos benthic marine algal flora prior to the Galapagos International Scientific Project, in *Proc. Symp. Galapagos Int. Scientific Project*, Bowman, R. I., Ed., University of California Press, Berkeley, 1966, 149.
407. Smith, S. V., Budget of calcium carbonate, southern California continental borderland, *J. Sediment. Petrol.*, 41, 798, 1971.
408. Smith, S. V., Production of calcium carbonate on the mainland shelf of southern California, *Limnol. Oceanogr.*, 17, 28, 1972.
409. Smith, S. V., Carbon dioxide dynamics: a record of organic carbon production, respiration, and calcification in the Eniwetok Reef flat community, *Limnol. Oceanogr.*, 18, 106, 1973.
410. Smith, S. V. and Kinsey, D. W., Calcium carbonate production, coral reef growth, and sea level change, *Science*, 194, 937, 1976.
411. Sneli, J. A, The Lithothamnion community in Nord-Möre, Norway, *Sarsia*, 31, 69, 1968.
412. Solms-Laubach, Graf zu, Die Corallinenalgen dis Golfes von Neapel und der angrenzended Meeresschnitte, *Fauna Flora Golfes Neapel*, 4, 1, 1881.
413. Sournia, A, Oxygen metabolism of a fringing reef in French Polynesia, *Helgol. Wiss. Meeresunters*, 28, 401, 1976.
414. South, G. R. and Adams, N. M., Marine algae of the Kaikoura Coast, a list of species, *Natl. Mus. N. Z., Misc, Ser. 1*, 1976.
415. Stearn, C. W., Scoffin, T. P., and Martindale, W., Calcium carbonate budget of a fringing reef on the west coast of Barbados. I. Zonation and productivity, *Bull. Mar. Sci.*, 27, 479, 1977.
416. Steneck, R. S. and Adey, W. H., The role of environment in control of morphology in *Lithophyllum congestum*, a Caribbean algal ridge builder, *Bot. Mar.*, 19, 197, 1976.
417. Stoddart, D. K., Ecology and morphology of recent coral reefs, *Biol. Rev.*, 44, 433, 1969.
418. Stosch, H. A. von, Kultureexperiment und Oekologie bei Algen, *Kiel. Meeresforsch.*, 18, 13, 1962.
419. Stosch, H. A. von, Wirkungen von Jod und Orsenit auf Meeresalgen in Kulture, in *Proc. Int. 4th Seaweed Symp.*, deVirville, A. D. and Feldman, J., Eds., Pergamon Press, Elmsford, N.Y., 142, 1964.
420. Stosch, H. A von, Observations on *Corallina, Jania* and other red algae in culture, in *Proc. Int. 6th Seaweed Symp.*, Margalef, R., Ed., Dirección General de Pesca Maritima, 389, 1969.
421. Suneson, S., Studien uber die Entwicklungsgeschichte der Corallinaceen, *Acta Univ. Lund., Sect. 2*, 33, 1, 1937.
422. Suneson, S., The structure, life-history and taxonomy of the Swedish Corallinaceae, *Acta Univ. Lund, Sect. 2*, 39, 1, 1943.
423. Suneson, S., Zur Spermatien bildung der Florideen, *Bot. Not.*, 1943, 373, 1943.
424. Suneson, S., Notes on *Schmitziella endophloea*, *K. Fysiogr. Sallsk. Lund. Forh.*, 14, 239, 1944.
425. Suneson, S., *Lithothamnion fornicatum* Fosl. ny for Sverige, *Bot. Not.*, 1944, 265, 1944.
426. Suneson, S., On the anatomy, cytology and reproduction of *Mastophora*, with a remark on the nuclear conditions in the spermatangia of the Corallinaceae, *K. Fysiogr. Sallsk. Lund. Forh.*, 15, 251, 1945.
427. Suneson, S., The cytology of the bispore formation in two species of *Lithophyllum* and the significance of the bispores in the Corallinaceae, *Bot. Not.*, 1950, 429, 1950.
428. Svedelius, N., Corallinaceae, in *Die naturlichen Pflanzenfamilien*, Nacträge zum Teil 1, t. 2, Engler, A. and Prantl, K., Eds., Leipzig, 1911, 257.
429. Tandy, G., Nomenclature of *Duthiea* Manza, *J. Bot. (London)*, 76, 115, 1938.
430. Taylor, W. R., Pacific marine algae of the Allan Hancock Expeditions to the Galapagos Islands, *Allan Hancock Pac. Exped.*, 12, 1, 1945.
431. Taylor, W. R., *Marine algae of the Northeastern coast of North America*, University of Michigan Press, Ann Arbor, 1957.
432. Taylor, W. R., *Marine Algae of the eastern tropical and subtropical coasts of the ericas*, University of Michigan Press, Ann Arbor, 1960.
433. Thuret, G. and Bornet, É., *Études phycologiques*, Masson, Paris, 1878.

434. **Tokida, J. and Masaki, T.,** Studies on the Melobesioideae of Japan. I, *Bull. Fac. Fish., Hokkaido Univ.,* 10, 83, 1959.
435. **Tokida, J. and Masaki, T.,** On the occurrence in Japan of a crustaceous coralline, *Polyporolithon, Bot. Mag.,* Tokyo, 73, 497, 1960.
435a. **Townsend, R. A.,** *Synarthrophyton,* a new genus of Corallinaceae (Cryptonemiales, Rhodophyta) from the southern hemisphere, *J. Phycol.,* 15, 251, 1979.
436. **Tsuda, R. T. and Randall, J. E.,** Food habits of the gastropods *Turbo argyrostoma* and *T. setesus,* reported as toxic from the tropical Pacific, *Micronesica,* 7, 153, 1971.
436a. **Turvey, J. R. and Simpson, P. R.,** Polysaccharides from *Corallina officinalis,* in, *Proc. 5th Int. Seaweed Symp.,* Young, E. G., and McLachlan, J. L., Eds., Pergamon Press, Elmsford, N.Y., 1966, 323.
437. **Van den Hoek, C.,** Algal vegetation-types along the open coasts of Curaçao, Netherlands Antilles. I, *Proc. Nederl. Akad. Wet. Ser C,* 72, 537, 1969.
438. **Van den Hoek, C.,** Algal vegetation-types along the open coasts of Curaçao, Netherlands Antilles. II. *Proc. Nederl. Akad. Wet. Ser. C,* 72, 559, 1969.
439. **Van den Hoek, C., Breeman, A. M., Bak, R. P. M., and Van Buurt, G.,** The distribution of algae, corals and gorgonians in relation to depth, light attenuation, water movement and grazing pressure in the fringing coral reef of Curaçao, Netherlands Antilles, *Aquatic Bot.,* 5, 1, 1978.
440. **Van den Hoek, C., Colijn, F., Cortel-Breeman, A. M., and Wanders, J. B. W.,** Algal vegetation typed along the shores of inner bays and lagoons of Curaçao, and of the lagoon Lac (Bonaire), Netherlands Antilles, *Verh. K. Akad. Wet. Afd. Natuurkd.,* 61, 1972.
441. **Wanders, J. B. W.,** The role of benthic algae in the shallow reef of Curaçao (Netherlands Antilles). I: primary productivity in the coral reef, *Aquatic Bot.,* 2, 235, 1976.
442. **Wanders, J. B. W.,** The role of benthic algae in the shallow reef of Curaçao (Netherlands Antilles). III: the significance of grazing, *Aquatic Bot.,* 3, 357, 1977.
443. **Weber, J. N. and Kaufman, J. W.,** Brucite in the calcareous alga *Goniolithon, Science,* 149, 996, 1965.
444. **Weber-van Bosse, and Foslie, M.,** The Corallinaceae of the Siboga-Expedition, *Siboga-Expeditie,* 61, 1904.
445. **Westbrook, M. A.,** Observations on nuclear structure in the Florideae, *Beih. Bot. Centralbl.,* 56A, 564, 1935.
446. **Woelkerling, W. J.,** *Mastophoropsis canaliculata* (Harvey in Hooker) gen. et comb. nov. (Corallinaceae, Rhodophyta) in southern Australia, *Br. Phyc. J.,* 13, 209, 1978.
447. **Womersley, H. B. S.,** The marine algae of Kangaroo Island. III. List of species, *Trans. R. Soc. South Aust.,* 73, 137, 1950.
448. **Womersley, H. B. S. and Bailey, A,** The marine algae of the Solomon Islands and their place in biotic reefs, *Phil. Trans. R. Soc., London, Ser., B,* 255, 433, 1969.
449. **Womersley, H. B. S. and Bailey, A.,** Marine algae of the Solomon Islands, *Phil. Trans. R. Soc., London, Ser. B,* 259, 257, 1970.
450. **Wray, J. L.,** Algae in reefs through time, *Proc. North. Am. Paleont. Conv.,* J, 1358, 1971.
451. **Wray, J. L.,** Late Paleozoic calcareous red algae, in *Fossil Algae, Recent Results and Development,* Flügel, E., Ed., Springer-Verlag, Berlin, 1977, 167.
452. **Wray, J. L.,** *Calcareous Algae,* Elsevier, Amsterdam, 1977.
453. **Yamanouchi, S.,** The life-history of *Corallina officinallis* var. *mediterranea, Bot. Gaz. Chicago,* 72, 90, 1921.
454. **Yendo, K.,** Corallinae verae of Port Renfrew, *Minn. Bot. Stud.,* 2, 711, 1902.
455. **Yendo, K.,** Corallinae verae Japonicae, *J. Coll. Sci. Imp. Univ. Tokyo,* 16, 1, 1902.
456. **Yendo, K.,** Enumeration of corallinaceous algae hitherto known from Japan, *Bot. Mag.,* Tokyo, 6, 185, 1902.
457. **Yendo, K.,** A study of the genicula of Corallinae, *J. l. Sci. Imp. Univ. Tokyo,* 19, 1, 1904.
458. **Yendo, K.,** Principle of systematizing Corallinae, *Bot. Mag.,* Tokyo, 19, 115, 1905.
459. **Yendo, K.,** A revised list of Corallinae, *J. Coll. Sci. Imp. Univ. Tokyo,* 20, 1, 1905.
460. **Yoneshigue-Braga, Y.,** Flora Marinha Bentonica da Baia de Guanabara e Cerrcanias. III. Rhodophyta. 2. Cryptonemiales, Gigartinales e rhodymeniales. *Inst. Pesquisas Mar,* Publ. No. 062, 1, 1972.
461. **Zaneveld, J. S.,** A *Lithothamnion* bank at Bonaire (Netherlands Antilles), *Blumea,* 4, 206, 1958.
462. **Yamada, Y.,** Sur la culture de quelques algues sur les fonds de sable ou de vase au Japan, *C.N.R.S.,* LXXXI, 251, 1959.
463. **Chemin, E.,** Le développement des spores chez les Rhodophycées, *Rev. Gen. Bot.,* 49, 205, 300, 353, 424, 478, 1937.
464. **Inoh, S.,** *Development of marine algae, kaiso nohassei,* Hokuryukan, Tokyo, 1947.
465. **Woelkerling, W. J.,** personal communication, 1979.

467. Pentecost, A., Calcification and photosynthesis in *Corallina officinalis* L. using the $_4CO_2$ method, *Br. Phycol. J.*, 13, 383, 1978.
468. Borowitzka, M. A. and Vesk, M., personal communication, 1979.
469. Webster, personal communication, 1972.
470. Cotton, A D., Marine algae. A Biological Survey of Clare Island ... , *Proc. R. Ir. Acad., Sect B*, 15, 1, 1912.
471. Van den Hoek, C., Cortrel-Breeman, A. M., and Wanders, J. B. W., Algal zonation in the fringing coral reef of Curaçao, Netherlands Antilles, in relation to zonation of corals and gorgonians, *Aquat. Bot.*, 1, 269, 1975.
472. Adams, N. M., The marine algae of the Wellington area. A list of species, *Dominion Museum*, 8, 43, 1972.
473. Lamb, I. M. and Zimmerman, M. H., Marine vegetation of Cape Ann, Essex County, Massachusetts. I. Seasonal succession. II. The occurring of the genus *Pantoneura* Kylin (Rhodophyta) in North America, *Rhodora*, 66, 217, 1964.
475. Saito, Y., Taniguchi, K., and Atobe, S., Phytosociological study of the intertidal marine algae. II. The algal communities on the vertical substratum faces on several directions, *Jpn. J. Ecol. (Nippon Seitai Gakkaishi)*, 20, 230, 1971.
476. Doty, M. S. and Newhouse, J., The distribution of marine algae into estuarine waters, *Am. J. Bot.*, 41, 508, 1954.
477. Littler, M. and Murray, S. N., Influence of domestic wastes on energetic pathways in rocky intertidal communities, *J. Appl. Ecol.*, 15, 583, 1978.
478. Earle, S. A., The influence of herbivores on the marine plants of Great Lameshur Bay, with an annotated list of plants, in Collete, B. B. and Earle, S. A., Results of the Tektite Program: Ecology of coral reef fishes. Natural History Museum, L. A. County, *Science Bull.*, 14, 17, 1972.
479. Dawson, E. Y., Marine red algae of Pacific Mexico. VII. Ceramiales: Ceramiaceae, Delesseriaceae, *Allan Hancock Pac. Exped.*, 26(1), 1962.
480. Chapman, V. J., Seaweeds and Their Uses, *Pitman Publ. Corp.*, New York, 1950.
481. Schwimmer, M. and Schwimmer, D., The role of algae and plankton in medicine. Grune and Stratton, New York, 1955.
482. Scagel, R. F., Marine plant resources of British Columbia, *Bull. Fish. Res. Bd. Can., Ottawa Bull.*, 127, 1961.
483. Riding, R., The Devonian genus *Keega* (Algae) reinterpreted as stromatoporoid basal layer, *Palaeontology*, 17, 565, 1974.
484. Pia, J., Florideae, In Himer, M., *Hanbuch der Palaeobotanik*, Druck and Verlag von R. Oldenbourg, Munich, 1, 96, 1927.
485. Elliot, G. F., Tertiary solenoporacean algae and the reproductive structures of the Structures of the Solenopoaceae, *Palaeontology*, 7, 695, 1965.
486. Wray, J. L., *Archaeolithophyllum*, an abundant calcareous alga in limestones of the Lansing Group (Pennsylvanian) Southeastern Kansas, *Bull. State Geol. Surv. Kans.*, 170(1), 1, 1964.
488. Johnston, G., A History of British Sponges and Lithophytes, W. H. Lizars, Edinburgh, 1842.
489. Bory, de Saint Vincent, Notice sur les Polypiers de la Grece, in Expedition scientific de Morée. Section des science physiques. Tome III. 1st Partie. Zoologie. Première section. Animaux vertebres, mollusques et polypiers. Chez F. G. Levrault, Paris, 1832, 204.
490. Batters, 1892.
491. Lindley, J., *The Vegetable Kingdom*, London, 1846.
492. Zanardini, G., Revista critica delle Corallinee, o Polipai calciferi de Lamouroux. *Atti R. 1st. Veneto Sci. Lett. Arti*, 3, 186, 1844.
493. Yamada, Y., Notes on some Japanese algae II. *J. Fac. Sci., Hokkaido Univ.*, V, 1(2), 65, 1931.
496. Kjellman, F. R., The algae of the Arctic Sea. A survey of the species, together with an exposition of the general characters and the development of the flora, *Kgl. Svensk Vetensk. Akad. Handl. Ser. 4*, 20(5), 1883.
497. Strömfelt, H. F. G., *Om algvegetationen vid Islands Kuster*, Göteborgs kngl. Vet. och Vitt. Samh. Handl., Ny. tid., 21, 2, 21, 1886.
498. DeToni, J. B., *Sylloge algarum, IV, Florideae*, Padua, 1905, p. 1523.
499. Esper, E. J. C, *Oie Pflanzenthiere in Abbildungen nebst Beschreibungen*, 3 vols., 2 Fortsetz, 2, p. 1798-1806 Nurnberg, 1788-1830.
500. Sonder, G., Nova Algarum genera et species, quas in itinere ad oras occidentales Novae Hollandiae, Collegit L. Preiss, Ph. Dr. *Bot. Zeit.*, 3(4), 49, 1845.

GLOSSARY

1. Adelphoparasite: A parasite that is closely related to its host; see alloparasite.
2. Adventitious: Arising secondarily from a position that is neither usual nor expected; not part of primary plant body.
3. Algal Ridge: A more or less elongate coralline structure at right angles to a wave front; the upper surface is at or near mean sea level in wave-exposed tropical areas; an incipient algal ridge is one that has not as yet reached the water surface.[18]
4. Alloparasite: A parasite that is not closely related to its host; see adelphoparasite.
5. Apical Meristem: A layer of medullary cells terminating filaments at branch or protuberance apices; overlain by epithallial cells in some taxa; cells capable of dividing transversely so as to increase the length of a branch.
6. Aragonite: Needle-like crystals of calcium carbonate; not formed by metabolic processes in coralline algae.
7. Arborescent: Bushy, shrubby, or treelike in habit, such as in some species of *Amphiroa*.
8. Atoll: A roughly ring-shaped system of reefs enclosing a lagoon and sometimes small islands.
9. Auxiliary Cell: A cell which receives a zygotic nucleus or its diploid derivative and then generates gonimoblast filaments.
10. Axial Conceptacle: A medullary conceptacle originating in line with and terminating the axis of a branch; in the Corallinoideae.
11. Axis: The main line of growth of an articulated frond from which branches arise, such as *Corallina*.
12. Bank: A reef system with the surface at least 6m below sea level.[395]
13. Barrier Reef: A reef system some distance from shore with an intervening lagoon usually more than 10m deep.
14. Basal Cell: A special cell bearing a supporting cell in female conceptacles or one or more spermatangial mother cells in male conceptacles; numerous of these cells in a conceptacle form a layer called the disc.
15. Bioherm: Any structural reef made up of living organisms and their carbonate remains.
16. Biolith: Same as bioherm.
17. Bisporangium: A sporangium containing two uninucleate or binucleate spores produced in asexual plants; see tetrasporangium.
18. Bispore: One of two haploid or diploid spores produced within a bisporangium; they may contain one or two nuclei.
19. Boiler: Same as cup reef.
20. Calcite: Plate-like crystals of calcium carbonate; contrast with aragonite.
21. Canal: The acellular space between conceptacular chamber and pore.
22. Carpogonial Filament: A two-celled filament consisting of a terminal carpogonium and a subterminal hypogynous cell; one to three borne on supporting cells in female conceptacle.
23. Carpogonium: A specialized oogonium made up of a base and an elongate process, the trichogyne.
24. Carposporangium: A diploid, single-celled sporangium terminating a gonimoblast filament in a female conceptacle.
25. Carpospore: A diploid spore produced singly within carposporangia in female conceptacles.

26. Carposporophyte: A parasitic plant developing after fertilization within a female conceptacle; consisting of the fusion cell(s) and associated gonimoblast filaments.
27. Cavity Cells: Comprising a tier of long cells breaking down during development so as to form a conceptacular chamber.
28. Chamber: In a conceptacle that space below the canal wherein are contained reproductive cells.
29. Coaxial Hypothallus: A multilayered hypothallus formed when divisions in the marginal meristem are synchronous so that the tissue consists of recumbent, curving tiers of cells.
30. Columella: A tuft of persistent cavity cells in the center of a tetrasporangial or bisporangial conceptacle.
31. Conceptacle: A uniporate or multiporate chamber housing reproductive cells surrounded by wall and roof.
32. Conceptacular Cap: Layers of cuticular and sometimes also cellular residue overlying a young conceptacle; usually sloughing off as the conceptacle matures.
33. Conceptacular Primordium: A group of meristematic cells destined to become a conceptacle and its contents.
34. Corniche: Same as bioherm.
35. Cortex: Tissue surrounding medulla in branches and protuberances; consisting of filaments ending at epithallial cells at branch surface; homologous to perithallus.
36. Cortical Conceptacle: A conceptacle originating in cortical tissue.
37. Cortical Meristem: A layer of cortical cells capable of dividing; each cell terminates a cortical filament and (usually) underlies an epithallial cell, hence this meristem is intercalary; growth results in increasing thickness of branch; homologous with perithallial meristem.
38. Cover Cell: Designating an epithallial cell; "deckzellen."
39. Crustose (Crustaceous): A crustlike plant growing appressed against the substratum.
40. Cumaphyte: A plant that thrives in turbulent water, as in the surf-zone.
41. Cup Reef: A small circular reef in which the edges reach the water surface; common in certain areas of the tropical Atlantic; also called boiler and microatoll.
42. Cuticle: Acellular and uncalcified layers bounding the outer surfaces of calcified structures.
43. Dichotomous Branching: Branched by repeated forkings in which branching parts give rise to two approximately equal branches.
44. Digitate: Handlike; compound, with parts arising from a common point, such as some branching intergenicula in *Corallina pinnatifolia*.
45. Dioecious: With male and female reproductive organs on separate plants.
46. Direct Secondary Pit-Connection: A pore and pit plug complex which forms between two cells when they grow together at one point, such as in the Lithophylloideae.
47. Disc (of Filaments): A compact group of special filaments which function to produce reproductive structures and associated cells inside a conceptacle, or which are so destined in a conceptacle primordium.
48. Distal: Opposite of basal or proximal; farthest from the point of attachment; at the free end.
49. Divaricate: Widely diverging or spreading; in, for example, dichotomous branching in *Jania adherens*.

50. Dorsiventral: With recognizable dorsal and ventral surfaces.
51. Endogenous: Formed internally; arising within a structure.
52. Endophyte: A plant growing within the tissues of another plant, the host, but not necessarily parasitically.
53. Endozoic (Endozootic): An organism living within the tissues of an animal, the host, but not necessarily parasitically.
54. Epilithic: Growing attached to a rock; saxicolous.
55. Epiphytic: Growing attached to a plant, the host, but not necessarily parasitically.
56. Epithallus: Outermost tissue of short cells (cover cells, cap cells) in one to six layers (rarely more) covering most calcified parts of thalli; produced outwardly from an intercalary meristem.
57. Excrescence: An outgrowth from the surface of a crustose coralline alga; see protuberance.
58. Fastigiate: When branches are nearly parallel and all point upward, such as *Jania verrucosa*.
59. Filament: A uniseriate chain of cells connected to one another by primary pit-connections.
60. Floridean Starch: An amylopectin storage product in red algae; organized into grains outside plastids.
61. Fringing Reef: An elongate reef system separated from shore by only a narrow body of water; same as bench bioherm.[22]
62. Frond: A complex of branches arising at one place from a holdfast in articulated coralline algae.
63. Fusion Cell: A large multinucleate cell formed following fertilization by fusion of numerous cells, including supporting cells, in base of female conceptacle; part of a carposporophyte from which gonimoblast filaments arise.
64. Gametophyte: A plant which is genetically constituted to produce gametes.
65. Geniculum: A group of completely or partly uncalcified cells located below every intergeniculum in articulated coralline algae.
66. Gonimoblast: A filament or cell complex formed within a female conceptacle after fertilization and resulting in the formation of carposporangia; see carposporophyte.
67. Hair: A colorless, elongate unicellular process; see trichocyte.
68. Hypogynous Cell: A cell subtending a carpogonium in a carpogonial filament.
69. Hypothallus: Lowermost tissue in crust in which one or several layers of filaments are oriented parallel to the substrate; homologous to medulla.
70. Intercalary Meristem: A meristem in which the dividing cells are subterminal, being overlain by epithallial cells; this type of meristem occurs in all calcified surfaces and, in some genera at crust margins and branch apices.
71. Intercalary Trichocyte: A trichocyte differentiated from either a marginal or an intercalary meristematic cell where the filament continues to grow while the trichocyte remains behind and becomes submarginal.
72. Intergeniculum: A calcified segment many of which make up the fronds of articulated coralline algae (except in *Yamadaea*, where a frond usually consists of a single intergeniculum); separated from one another by uncalcified genicula.
73. Lateral Conceptacle: A cortical conceptacle originating at the surface of a flattened or cylindrical intergeniculum, such as in *Bossiella*.
74. Lobe: An upward projecting extension of an intergeniculum, such as in *Cheilosporum*.

75. Marl (Maerl in French): A shallow-water sediment in which more than half of the material consists of small, unattached, irregularly branched, nonarticulated plants (contrast with rhodolith).
76. Marginal Conceptacle: A medullary conceptacle originating in the margin of a flattened intergeniculum; in the Corallinoideae.
77. Marginal Meristem: Cells terminating hypothallial filaments at the margin of a crust; the cells are overlain by epithallial cells in some taxa; growth results in increasing extension of a crust.
78. Medulla: Tissue consisting of core of filaments within branches, whether articulated or not; homologous to hypothallus.
79. Medullary Conceptacle: A conceptacle in the Corallinoideae originating in medullary tissue.
80. Megacell: An enlarged cell formed by differentiation from a trichocyte; megacells often persist after burial by surrounding growth and are distinctive by being larger than neighboring cells.
81. Meristem: A layer or group of cells wherein cell division occurs.
82. Micratoll: Same as cup reef.
83. Minisegment: One of several small pads of calcified tissue attached to genicula in *Amphiroa bowerbankii*.
84. Monoecious: With male and female reproductive organs on the same plant.
85. Multiaxial: A type of thallus construction with many axial filaments, each terminating in an apical cell.
86. Nullipore: An obsolete name for nonarticulated coralline algae that seemingly lacked pores; coined to distinguish these plants from coral animals.
87. Palmate Branching: Several intergenicula radiating fanlike or fingerlike from a subtending intergeniculum.
88. Paraphysis: A sterile filament among the reproductive cells in a conceptacle.
89. Percurrent Axis: A row of axial intergenicula and genicula extending from base to apex of a frond; the intergenicula may be pinnately branched.
90. Perithallial Meristem: A layer of perithallial cells capable of dividing; each cell terminates a perithallial filament and (usually) underlies an epithallial cell, hence, this meristem is intercalary; growth results in increasing thickness of crust; homologous with cortical meristem.
91. Perithallus: Tissue arising from hypothallus; growth is by the activity of a meristem just below an overlying epithallus; perithallial filaments are approximately perpendicular to the adjacent hypothallus; homologous with cortex.
92. Pinnate branching: A percurrent axis with lateral branches arising one from each side of compressed axial intergenicula; three genicula and their surmounting intergenicular emanate from each branching intergeniculum with the central one dominating and continuing the main axis.
93. Pit Body: A small subspherical body appressed to a primary pit-connection plug and extending into the cell; also called pit cap.
94. Pit-Connection: See Primary Pit—Connection.
95. Pit Plug: A dense structure coming to occupy the space left during cytokinesis when the new cell walls do not form a complete closure; primary pit plugs usually form on the central axis of a filament; sometimes bracketed by pit bodies.
96. Pore: The opening of the conceptacular canal to the outside of the conceptacle.
97. Pore Cells: Cells surrounding pore plugs in some Melobesioideae; staining intensely with phosphotungstic hematoxylin.[8]

98. Pore Plug: In the Melobesioideae an acellular gelatinous material filling the canal above a tetrasporangium or bisporangium; expelled prior to the time sporangia (or spores) pass through the canal; also called sporangial plug.
99. Primary Branching: Branch origin being a group of cells in apical meristem in articulated coralline algae; contrast with secondary branching.
100. Primary Meristem: A meristem made up of cells terminating hypothallial or medullary filaments; in many genera marginal and apical meristems are primary.
101. Primary Pit-Connection: A pore and pit plug complex forming in the walls between two cells when they divide from one.
102. Procarp: A carpogonium and auxiliary cell complex when both structures are members of a common branch system.
103. Protuberance: A simple or branched outgrowth from the surface of a crustose coralline algae; see excrescence.
104. Pseudodichotomous Branching: That branching of a filament accomplished when a meristematic cell obliquely cuts off a new cell and then transversely another cell so as to form new meristematic cells.[62]
105. Pseudolateral conceptacle: Conceptacles originating in medullary tissue in apex of aborted secondary branch in *Corallina;* these conceptacles protrude more than do lateral conceptacles.
106. Rhizoid: An anchoring structure consisting of a single elongate cell or process.
107. Rhizome: An articulated branch growing from basal intergenicula and serving to secure frond to substrate.
108. Rhodolith: An unattached nodular, coralline structure sometimes occurring in large numbers to a depth of 200 m; usually formed around a nucleus and made up of concentric layers of crustose coralline algae, with or without protuberances (contrast with marl).
109. Secondary Branching: Branch origin being a group of cells in cortical meristem in articulated coralline algae; originating after primary growth in the area has ceased; contrast with primary branching.
110. Secondary Hypothallus: A hypothallus originating *de novo* from mature perithallial or hypothallial cells.
111. Secondary Pit-Connection: A pit-connection developed between two adjacent cells.
112. Secondary Structures: Comprised of filaments produced by renewed and rapid growth of perithallial or cortical meristems, e.g., secondary hypothallia, secondary cortices, secondary conceptacles, and secondary branches.
113. Sorus (Sori): A group or cluster of reproductive cells.
114. Spermatangial Mother Cell: A special spermatangium-producing cell developing from vegetative or basal cells in male conceptacles.
115. Spermatangium: One of numerous cells produced from spermatangial mother cells in the fertile layer of a male conceptacle; spermatia may be extruded from spermatangia.
116. Spermatium: A male gamete extruded or budded from a spermatangium in red algae.
117. Sporangial Plug: Same as Pore Plug.
117a. Sporophyte: A spore-producing phase of an algal life history, alternating with the gametophyte generation; usually diploid.
118. Staining Body: A small granule, possible proteinaceous, occurring in large numbers in vegetative cells in certain species of nonarticulated coralline algae in *Phymatolithon*.[3]
119. Stalk Cell: A cell subtending a tetrasporangium or bisporangium.

120. Supporting Cell: A cell bearing one or more carpogonial filaments; these cells constitute a layer in the disc of a female conceptacle.
121. Terminal Trichocyte: A trichocyte differentiated from a marginal meristematic cell; a terminal trichocyte terminates a filament and subsequently becomes submarginal by the growth of neighboring filaments; present in some species of *Fosliella*.[86]
122. Tetrasporangium: A zonate sporangium containing four haploid tetraspore; produced in tetrasporophytes.
123. Tetraspore: One of four haploid spores produced within a tetrasporangium.
124. Tetrasporophyte: A plant that produces tetraspores by meiotic divisions in tetrasporangia.
125. Transfer Tube: A nonseptate connection growing from the base of a fertilized carpogonium to the supporting cell in the same filament system.
126. Trichocyte: A complex of one to three uncalcified cells bearing a hair which persists for varying lengths of time.
127. Trichogyne: An elongated apical process of a carpogonium projecting through the conceptacle pore at maturity; fusing with passing spermatangia (or spermatia).
128. Whorl: A group of three or more similar branches originating at the same level in a frond.
129. Wing: A flat lateral protrusion of an intergeniculum, such as in *Bossiella*.

APPENDIX 1

GENERA IN THE CORALLINACEAE

The following is an annotated list of the genera of coralline algae currently recognized. The principal characteristics that are listed include only those necessary to gain an idea of how a genus is distinctive. The diagnostic characteristics of the subfamilies are in Tables 3 and 4, Chapter 1. Dr. P. C. Silva's compilation for the *Index Genericorum* was the source of some of these data.

Alatocladia (Yendo) Johansen.[223] Type Species: *A. modesta* (Yendo) Johansen.[223] *Cheilosporum anceps* var. *modestum* Yendo.[455]

Main Characteristics:
—Conceptacles axial and marginal in origin
—Intergenicular medullary filaments interlacing
—Branching irregular
—Pores of tetrasporangial conceptacles opening on surfaces of the flat intergenicula.

Main References: Segawa;[388] Johansen,[223] Murata and Masaki.[326]

Comments: *Alatocladia* is a poorly known monotypic genus restricted to Japan. It is probably related to *Calliarthron*, with which it shares the interlacing filament characteristic. Unlike *Calliarthron*, however, *Alatocladia* lacks cortical conceptacles. Only tetrasporangial plants of the single species have been recorded.[388]

Amphiroa Lamouroux.[255] Type Species: *A. tribulus* (Ellis et Solander) Lamouroux.[255] *Corallina tribulus* Ellis et Solander.[156]

Main Characteristics:
—One or more tiers of genicular cells
—Uncalcified cortical cells part of genicula
—Cell lengths in genicula same as in intergenicula
—Several tiers of medullary cells per intergeniculum
—Some species modified and growing attached to crustose coralline algae, others with crustose holdfasts.

Main References: Suneson,[421] Segawa,[377-378] Johansen,[222-224] Ganesan,[175] and Cabioch.[69,76]

Antarcticophyllum (Lemoine) Mendoza.[318] Type Species: *A. aequabile* (Foslie) Mendoza.[318] *Lithophyllum discoideum* f. *aequabile* Foslie.[162]

Main Characteristics:
—Hypothallus thin
—Perithallus thick
—Epithallus multi-layered

Main Reference: Mendoza.[318]

Comments: It is not clear to which of the two melobesioidean tribes this genus belongs. Mendoza[318] also included *A. subantarcticum* in this genus.

Arthrocardia Decaisne.[128] Lectotype species: *A. corymbosa* (Lamarck) Decaisne.[128] *Corallina corymbosa* Lamarck.[254]

Main Characteristics:
—Conceptacles axial in origin
—Fertile intergenicula usually bearing two branches
—Branching pinnate in vegetative parts of fronds

Main References: Manza,[294,300-301] Segawa,[388] Ganesan,[174] and Johansen.[223,227]

Comments: Although most species in this genus occur in South Africa, it is also present in India,[174] Brazil,[243] and the eastern Pacific Ocean.[227] Its presence in Australia is doubtful. Probably related to *Corallina,* it differs in being relatively more robust and in having deeply embedded conceptacles in intergenicula which consistently branch dichotomously. Furthermore, *Arthrocardia* has fusion cells giving rise to carposporangial filaments over the entire upper surface,[174] and male conceptacles with elongate beaks.

Bossiella Silva.[405] Type Species: *B. plumosa* (Manza) Silva.[405] *Bossea plumosa* Manza.[294]

Main Characteristics:
— Conceptacles solely lateral in origin
— Subgenus *Bossiella:* intergenicula flat; four species
— Subgenus *Pachyarthron:* (Manza) Johansen;[223] intergenicula terete

Main References: Manza,[294-295,301] Segawa,[381] Johansen,[223,226-228,230] and Ganesan.[176]

Comments: Manza[294] first placed articulated species with flat intergenicula and lateral conceptacles in a separate genus; he suggested *Bossea*. However, this name had previously been used and hence Silva[405] suggested *Bossiella*. Based mostly on the results of Segawa's[381] study of *Amphiroa cretacea* (= *B. cretacea*) the monotypic *Pachyarthron* was later included as a subgenus in *Bossiella*.[223] *Bossiella* is almost wholly restricted to the Pacific Ocean, with only a few records from the eastern coast of Argentina.

Calliarthron Manza.[294] Type Species: *C. cheilosporioides* Manza.[294]

Main Characteristics:
— Conceptacles marginal and cortical in origin
— Medullary filaments interlacing

Main References: Manza,[294-295,301] Segawa,[383] Ganesan,[176] and Johansen.[223]

Comments: Present on both sides of the Pacific Ocean, this genus is probably related to the two Japanese genera *Alatocladia* (both have interlaced medullary filaments) and *Marginisporum* (both have marginal and cortical conceptacles).

Chaetolithon Foslie.[160d] Type Species: *C. deformans* (Solms-Laubach) Foslie.[160d] *Melobesia deformans* Solms-Laubach.[412]

Main Characteristics:
— parasitic on *Jania* sp.
— Conceptacles multiporate

Main References: Solms-Laubach;[412] Kylin.[253]

Comments: Very little is known about this monotypic genus from Australia. Kylin's[256] terse description suggests that it resembles *Choreonema* except that the tetrasporangial conceptacles are multiporate.

Cheilosporum (Decaisne) Zanardini.[492] Lectotype Species: *C. sagittatum* (Lamouroux) Areschoug.[32] *Corallina sagittata* Lamouroux;[257] (Schmitz.)[370]

Main Characteristics:
— Conceptacles marginal
— Intergenicular lobes usually conspicuous, apiculate or rounded
— Main branching dichotomous

Main References: Johansen.[223,233]

Chiharaea Johansen.[221] Type Species: *C. bodegensis* Johansen.[221]

Main Characteristics:
— Conceptacles axial, one to three per intergeniculum
— 12 or fewer intergenicula per frond
— Intergenicula compressed, recumbent
— Pores of tetrasporangial and carposporangial conceptacles dorsal on intergenicular surfaces
— Crustose base extensive

Main References: Johansen;[221,226] Lebednik.[261]

Comments: This monotypic genus seems to be admirably adapted to living in violent surf. It has been located only on the west coast of North America.

Choreonema Schmitz.[370] Type Species: *C. thuretii* (Bornet) Schmitz.[370] *Melobesia thuretii* Bornet.[433]

Main Characteristics:
—Parasitic on species of *Jania, Haliptilon,* and *Cheilosporum*
—Vegetative parts consisting of unpigmented filaments among host filaments
—conceptacles growing on surfaces of host intergenicula

Main References: Suneson,[421,423] Dawson,[116-117] and Cabioch.[76]

Comments: Only one species has been included in this genus.

Clathromorphum Foslie.[160b-160c] Lectotype Species: *C. compactum* (Kjellman) Foslie.[160c] *Lithothamnium compactum* Kjellman;[496] (Mason.[309])

Main Characteristics:
—Epithallus two to several cell layers thick.

Main References: Adey,[4-5] Adey and Adey,[20] Adey and Johansen,[23] and Lebednik.[261-262]

Comments: Cabioch[76] considered this a subgenus under *Lithothamnium*.

Corallina Linnaeus.[280] Lectotype Species: *C. officinalis* Linnaeus,[280] (Schmitz.[370])

Main Characteristics:
—Conceptacles axial in origin
—Branching pinnate

Main References: Suneson,[421] Ganesan,[175-177] Johansen,[223,225] and Cabioch.[76]

Comments: *Corallina officinalis* is the first articulated species validly published.[280] Many species of *Corallina* have been described, and it is impossible to say how many represent forms rather than species in the current sense. Specimens resembling *C. officinalis* are present in both the northern and southern hemispheres, but absent in the tropics.

Dermatolithon Foslie.[160d] Type Species: *D. pustulatum* (Lamouroux) Foslie.[160d] *Melobesia pustulata* Lamouroux.[256]

Main Characteristics:
—crusts comprised of one or more tiers of elongate cells

Main References: Suneson,[421,423,427] Dawson,[114] Tokida and Masaki,[434] Ganesan,[171] Huvé,[216] Masaki,[303] Lemoine,[273-274] Cabioch,[76] and Chamberlain.[86-87]

Comments: Adey[13] removed several species generally referred to *Dermatolithon* to *Tenarea* and does not recognize the former genus. On the other hand, Cabioch[74,76] recognized both genera, with *Dermatolithon* reserved for simple (possibly primitive) species and *Tenarea* for those that have the more complex organization described in Chapter 3.

Ezo Adey, Masaki et Akioka.[25] Type Species: *E. epiyessoense* Adey, Masaki et Akioka.[25]

Main Characteristics:
—achlorophyllous
—Thalli consisting only of small vegetative lobes, short haustoria, and conceptacles
—Parasitic on *Lithophyllum*

Main Reference: Adey, et al.[25]

Comments: Like *Kvaleya,* this monotypic genus is an excellent example of an adelphoparasite.

Fosliella Howe.[213] Type Species: *F. farinosa* (Lamouroux) Howe.[213] *Melobesia farinosa* Lamouroux.[256]

Main Characteristics:
—Crusts thin, perithallus sometimes absent except around conceptacles
—Trichocytes present

Main References: Suneson,[421,423] Balakrishnan,[35] Mason,[309] Dawson,[116-117] Ganesan,[172] Masaki,[303] Adey,[13] Cabioch,[76] Adey and Adey,[20] Bressan,[57] and Chamberlain.[84-85]

Comments: Since 1920 most phycologists have recognized *Fosliella* for thin, trichocyte-containing crusts with uniporate tetrasporangial conceptacles. Dawson[116-117] restricted the genus to epiphytic species, but Adey and Adey[20] showed that epilithic species are also present. Balakrishnan,[35] Kylin,[253] and Masaki and Tokida[304-305] considered the species usually assigned to *Fosliella* under *Melobesia*, although Masaki[303] recognized *Fosliella*.

Goniolithon Foslie[160c] Type Species: *G. papillosum* (Zanardini ex Hauck) Foslie,[160c] *Lithothamnium papillosum* Zanardini ex Hauck.[201]

Main Characteristics:
—Thalli thick, made up of aggregated *Dermatolithon*-like crusts in tiers
—Thalli lacking primary hypothallial tissues which have been replaced by secondary hypothallial tissues ("faux-hypothalle")[76]
—Most cells, except epithallia, more than three times as long as broad

Main References: Setchell and Mason;[399] Cabioch.[71,76]

Comments: *Goniolithon*, with two subgenera, *Eugoniolithon* and *Cladolithon*, the latter for fruticulose species, was described with a very brief description by Foslie.[160c] The type and only species under *Eugoniolithon* was *G. papillosum* (as described by Hauck) and this should be considered the type species of the genus.[399] *Cladolithon* was a questionable taxon and included only *G. byssoides*, a heterogeneous assemblage of plants.

Later in the same year,[160d] Foslie changed the name *Eugoniolithon* to subgenus *Lepidomorphum* and included *G. papillosum* among its four species. Under *Cladolithon* he listed 18 species, with *B. byssoides* first and *G. moluccense* second. In 1900, Foslie[161] rejected *Goniolithon* as he earlier construed it and characterized his new concept by tetrasporangia distributed over the floor of the conceptacle and (probably?) with scattered megacells. According to Setchell and Mason,[399] this new concept was based on *G. moluccense*, which he had included previously under the subgenus *Cladolithon*. This second concept of *Goniolithon* was clarified further in 1904[161b] and 1909.[164] Because of the fact that two distinct concepts of *Goniolithon* existed, Setchell and Mason[399] suggested that the second should be segregated as a new genus, *Neogoniolithon*, with *N. fosliei* as its type species. *Goniolithon*, with *G. papillosum* as its type species, should still be recognized.

Haliptilon (Decaisne) Lindley.[491] Type Species: *H. cuvieri* (Lamouroux) Johansen et Silva.[236] *Corallina cuvieri* Lamouroux.[256]

Main Characteristics:
—Carposporophytic fusion cells up to 35 μm thick and less than 130 μm wide
—Chamber of male conceptacles 90-250 μm in diameter
—Fewer than 15 mature tetrasporangia per conceptacle
—Branches arising from tetrasporangial and carposporangial conceptacles, not from male conceptacles

Main References: Johansen;[225-226] Johansen and Silva.[236]

Comments: For more than a century the species now placed in *Haliptilon* have been recognized as constituting a distinctive group of taxa.[129] Because of their vegetative similarity to *Corallina*, however, they have usually been placed in this genus, even though Lindley[491] recognized the genus *Haliptilon*. The reproductive structures are more similar to those of *Jania* than to *Corallina*, and recently[236] *Haliptilon*, *Jania*, and *Cheilosporum* were segregated into the tribe Janieae. In all probability several other species presently included in *Corallina* should be removed to *Haliptilon*.

Heteroderma Foslie.[164] Lectotype Species: *H. subtilissima* (Foslie) Foslie.[164] *Melobesia subtilissima* Foslie;[161b] (Setchell and Mason.[399])

Main Characteristics:
—Crusts thin, epiphytic
—Trichocytes absent

Main References: Suneson,[421,423] Mason,[309] Dawson,[116-117] Masaki and Tokida,[308] Masaki,[303] Cabioch.[76]

Comments: From the time the name *Heteroderma* was first coined as a subgenus under *Choreonema*,[160c] its circumscription was revised several times and even recently it has been considered a genus (as it is here) or a subgenus under *Fosliella*.[76] Mason[309] reviewed the nomenclatural history of this genus. Foslie,[161] after changing his concept of *Choreonema* to include only one species, had *Heteroderma* and *Eumelobesia* as subgenera under his large genus *Melobesia*. The former subgenus contained plants with several cell layers, and the latter plants only one cell layer thick. However, a few years later Foslie[162] differentiated the subgenera *Heteroderma* and *Eumelobesia* on the lack of heterocysts (trichocytes) in the former and their presence in the latter. In 1908[163], he transferred those species of subgenus *Heteroderma* having oligostromatic thalli into a new subgenus, *Pliostroma*. *Heteroderma* was finally accorded generic status[164] with two subgenera, *Heteroderma* and *Pliostroma* as they had been designated when still under *Melobesia*. With the establishment of the genus *Heteroderma*, and the transfer of most of the species out of *Melobesia*, the latter genus became monotypic, containing only *M. farinosa*. (This species was recognized as *Fosliella*,[213] see comments under *Fosliella*.) For now, *Heteroderma* will be treated as a genus lacking subgenera and closely related to *Fosliella*.

Hydrolithon (Foslie) Foslie.[164] Lectotype Species: *H. reinboldii* (Weber-van Bosse et Foslie) Foslie.[164] *Lithophyllum reinboldii* Weber-van Bosse et Foslie;[444] (Mason.[309])

Main Characteristics:
—Hypothallus monostromatic
—Megacells single

Main References: Mason,[309] Dawson,[116-117] Adey,[13] and Gordon, et al.[190]

Comments: Whether *Hydrolithon* should be considered a distinct genus or not needs to be studied. Cabioch[76] expressed uncertainty and excluded it from her classification scheme; she stated that it should perhaps be considered a subgenus under *Porolithon* or *Neogoniolithon*. In most descriptions (e.g., Adey;[13] Johansen,[232]) it is characterized by single megacells and a unistratose hypothallus, the latter feature unlike the situation in *Porolithon* and *Neogoniolithon*.

Jania Lamouroux.[255] Lectotype Species: *J. rubens* (Linnaeus) Lamouroux.[256] *Corallina rubens* Linnaeus;[280] (Manza.[294])

Main Characteristics:
—Main branching dichotomous
—Conceptacles axial
—Fertile intergenicula in all but male plants branching dichotomously

Main References: Suneson,[421] Dawson,[113] Ganesan,[173] Cabioch,[63,76] and Johansen.[223,225-226]

Kvaleya Adey et Sperapani.[28] Type species: *K. epilaeve* Adey et Sperapani.[28]

Main Characteristics:
—achlorophyllous
—parasitic on *Leptophytum*

Main Reference: Adey and Sperapani.[28]

Comments: *Kvaleya* is a monotypic genus.

Leptophytum Adey.[8] Type Species: *L. laeve* (Strömfelt) Adey.[8] *Lithophyllum laeve* Strömfelt.[497]

Main Characteristics:
—Crustose, thin

—Epithallial cells large, rounded
—Hypothallus multistratose, not coaxial

Main References: Adey;[8] Adey and Adey.[20]

Comments: Cabioch[76] indicated that she would consider *Leptophytum* a subgenus under *Lithothamnium*.

Litholepis Foslie.[162] Lectotype Species: *L. caspica* (Foslie) Foslie.[162] *Melobesia caspica* Foslie;[161] (Hamel and Lemoine.[197])

Main Characteristics:
—Crusts thin, perithallus often lacking
—Overgrowing one another
—Epithallus absent

Main References: Dawson,[116-117] Cabioch,[72] and Adey.[13]

Comments: *Litholepis* is poorly known. Adey[13] treated it with *Lithoporella* suggesting that later study may clarify the limits of these genera, as well as their relationships to *Heteroderma*. Cabioch[76] stated that in *Litholepis* epithallia are absent and that it might best be considered a subgenus under *Fosliella*.

Lithophyllum Philippi.[353] Lectotype Species: *L. incrustans* Philippi;[353] (Foslie.[160c])

Main Characteristics:
—Hypothallus uni or multistratose
—Perithallus usually copious

Main References: Suneson,[421,423] Dawson,[116-117] Masaki and Tokida,[308] Adey,[7] Masaki,[303] Cabioch,[66-67,76] Steneck and Adey,[416] Giraud and Cabioch.[185]

Lithoporella (Foslie) Foslie,[164]. Lectotype Species: *L. melobesioides* (Foslie) Foslie.[164] *Mastophora melobesioides* Foslie, (Kylin,[253]).

Main Characteristics:
—Vegetative parts lacking epithallia
—Crusts tending to overgrow one another

Main References: Segawa,[390] Masaki,[303] Lemoine,[273-275] and Cabioch.[76]

Comments: Adey[13] considered *Lithoporella* and *Litholepis* inseparable on the basis of current knowledge, although he thought that with study they, as well as *Heteroderma*, might be satisfactorily defined. *Lithoporella* has an overlapping habit of growth and cells generally larger than those in other extant coralline algae. It was among the last genera that Foslie described and has not had a confused history of circumscription.

Lithothamnium Philippi.[353] Lectotype Species: *L. ramulosum* Philippi;[353] (Mason.[309])

Main Characteristics:
—Crustose, some with protuberances somemarl-formers
—Epithallial cells with angular "eared" peripheral cell walls
—Hypothallus multistratose, not coaxial
—Spermatangia produced in dendroid branching filaments

Main References: Suneson,[421] Mason,[309] Dawson,[116-117] Adey,[8] Masaki,[303] and Cabioch.[76]

Comments: Mason[309] discussed the nomenclatural background of *Lithothamnium*. Both this genus and *Lithophyllum* were first described by Philippi,[353] the former to include coralline algae with crustose thalli, and the latter those with foliose thalli. He assigned five species to *Lithothamnium*, some of which are now placed in other genera, with the last of the five, *L. ramulosum*, chosen as lectotype species by Mason.[309] This species has often incorrectly been combined under *L. fruticulosum*, a later species; Adey[13] synonymized this latter species under *L. ramulosum*. The choice of *L. fasciculatum* as lectotype[370] is incorrect because it was not included in Philippi's original generic presentation.[309] The type of *L. ramulosum* has not been found.[13] *Lithothamnium calcareum,* now placed in *Phymatolithon,*[3] has also been given as the type species of *Lithothamnium.*[253]

In the present century, *Lithothamnium* and *Lithophyllum* have often been segregated on the basis of multipored conceptacles in the former and single-pored ones in the latter, features first recognized by Heydrich.[203] A current description of *Lithothamnium* was given by Adey,[8] who first described the distinctive epithallial cells and contrasted them with those in *Phymatolithon* and other melobesioid genera.

Lithothrix Gray.[192] Type Species: *L. aspergillum* Gray.[192]
Main Characteristics:
— Single tier of long genicular cells
— No cortical cells from genicula
— Ratio of genicular to intergenicular cell length as much as 40
— One medullary cell tier of short cells per intergeniculum
— Cells in crustose base resembling those in *Dermatolithon*

Main References: Segawa,[387] Johansen,[223] Ganesan and Desikachary,[180] Gittins,[187] and Gittins and Dixon.[188]

Marginisporum (Yendo) Ganesan.[175] Type Species: *M. crassissimum* (Yendo) Ganesan.[175] *Amphiroa crassissima* Yendo.[455]
Main Characteristics:
— Conceptacles marginal and lateral

Main References: Ganesan;[175] Johansen.[223]

Mastophora Decaisne.[128] Type Species: *M. licheniformis* Decaisne.[128]
Main Characteristics:
— Ribbon coralline, attached at one end
— Thalli thin, perithallus lacking

Main References: Setchell;[396] Cabioch.[76]

Comments: Setchell[396] analyzed the origin of the name *Mastophora* and found that Decaisne[128] properly gave a Latin diagnoses, listed a type species, and illustrated it. Confusion resulted when, later in the same year,[129] he reduced the genus to the status of a section under *Melobesia.* Some subsequent workers treated Decaisne's two papers as one and further confused the names *M. licheniformis* and *Mesophyllum lichenoides* (Areschoug,[132] Kützing,[251] and DeToni,[498]).

Mastophoropsis Woelkerling.[446] Type Species: *M. canaliculata* (Harvey in Hooker) Woelkerling.[446]
Main Characteristics:
— Ribbon coralline, attached at one end
— Tetrasporangial conceptacles multiporate

Main Reference: Woelkerling.[446]

Melobesia Lamouroux.[255] Lectotype Species: *M. membranacea* (Esper) Lamouroux.[255] *Corallina membranacea* Esper;[499] (Howe.[213])
Main Characteristics:
— Thin crust, usually epiphytic
— Tetrasporangial conceptacles multiporate

Main References: Suneson,[421] Lee,[267] Cabioch,[76] and Lebednik.[263]

Mesophyllum Lemoine.[270] Lectotype Species: *M. lichenoides* (Ellis et Solander) Lemoine.[270] *Millepora lichenoides* Ellis et Solander;[156] (Hamel and Lemoine.[197])
Main Characteristics:
— Hypothallus coaxial
— Conceptacle cap containing only epithallial cells

Main References: Suneson,[421] Masaki,[303] Adey and Johansen,[23] Cabioch,[71] Adey and Adey,[20] and Lebednik.[265]

Metagoniolithon Weber-van Bosse.[444] Lectotype Species: *M. charoides* (Lamouroux) Weber-van Bosse.[444] *Amphiroa charoides* Lamouroux.[256]
Main Characteristics:
—Genicula multi-tiered and producing branches
—Mucilaginous caps over branch apices
Main References: Johansen,[223] Cabioch,[74-76] Ganesan,[178] and Ducker.[144]
Metamastophora Setchell[396] Type Species: *M. flabellata* (Sonder) Setchell.[399-400] *Melobesia flabellata* Sonder.[500]
Main Characteristics:
—Ribbon coralline, attached at one end
—Perithallus present
Main References: Setchell,[396] Suneson,[426] Papenfuss,[347] Cabioch,[76] and Woelkerling.[465]
Neogoniolithon Setchell et Mason.[400] Type Species: *N. fosliei* (Heydrich) Setchell et Mason.[400] *Lithothamnium fosliei* Heydrich.[203]
Main Characteristics
—Megacells single or in single vertical series
Main References: Lee,[266] Masaki,[303] Cabioch,[65,76] and Adey.[13]
Comments: The erection of *Neogoniolithon* by Setchell and Mason[400] was mentioned in the comments under *Goniolithon*. *Neogoniolithon* is closely related to the poorly circumscribed *Hydrolithon* and *Porolithon*. According to published descriptions, *Neogoniolithon* and *Porolithon* would appear to be clearly segregated on the grouping of megacells. However, Cabioch[76] claimed that she finds intermediate forms that cloud the apparent distinctiveness of these two genera. The trichocytes and megacells in these genera are described in Chapter 2.
Phymatolithon Foslie.[160b] Type Species: *P. polymorphum* (Linnaeus) Foslie.[160c] *Corallina polymorpha* Linnaeus.[281]
Main Characteristics:
—Crusts or marl
—Epithallial cells rounded
Main References: Suneson,[421] Adey,[3,6-7] Adey and McKibbin,[27] Cabioch,[76] Adey and Adey.[20]
Comments: Cabioch[76] considered this a subgenus under *Lithothamnium*.
Porolithon (Foslie) Foslie.[164] Lectotype Species: *P. onkodes* (Heydrich) Foslie.[164] 1909b. *Lithophyllum onkodes* Heydrich.[203]
Main Characteristics:
—Megacells in horizontally oriented strata
—Hypothallus multistratose
Main References: Dawson,[117] Desikachary and Ganesan,[133] Lee,[266] Masaki,[303] Adey,[13] and Cabioch.[70-76]
Schmitziella Bornet et Batters in Batters.[490] Type Species: *S. endophloea* Bornet et Batters in Batters.[490]
Main Characteristics:
—Growing between the cell wall layers of *Cladophora*
—Producing bisporangia in nemathecia which erupt through the host cell wall
Main References: Suneson,[424] Chapman,[89] and Cabioch.[76]
Serraticardia (Yendo) Silva.[405] Type Species: *S. maxima* (Yendo) Silva.[405] *Cheilosporum maximum* Yendo.[455]
Main Characteristics:
—Conceptacles axial and lateral in origin
—Branching pinnate
Main References: Segawa,[384] Silva,[405] Ganesan,[176] and Johansen.[223,226]

Comments: Two species have been placed in this genus, one from the western Pacific and the other from the eastern Pacific. The fact that the plants contain axial as well as lateral conceptacles suggests that it is intermediate between *Corallina* and *Bossiella*. The genus is similar to *Corallina* in that the branching is densely pinnate throughout. Certain forms of *Serraticardia macmillanii* may be difficult to distinguish from *Bossiella plumosa* and others may be difficult to distinguish from *Corallina officinalis* var. *chilensis* from the coast of California.[226]

Sporolithon Heydrich.[203] Type Species: *S. ptychoides* Heydrich.[203]

Main Characteristics:
—Tetrasporangia in individual conceptacles usually arranged in horizontal lenses

Main References: Denizot,[132] Womersley and Bailey,[449] Cabioch,[72] and Gordon, et al.[190]

Comments: This unusual and supposedly primitive genus is the only coralline taxon where cruciately divided sporangia have been reported.[132,449] As Denizot[132] has suggested, *Sporolithon* may be considered as an aberrant genus in the Corallinaceae, and perhaps it should be removed from the family.[449] At the present, too little information is available to consider this problem adequately. *Sporolithon* produces sporangia in individual chambers, fields of which are sometimes called nemathecia. These features, plus the possible lack of epithallus,[449] indicate that further studies of this genus are sorely needed.

Synarthrophyton Townsend.[435a] Type Species: *S. patena* (Hooker fils et Harvey) Townsend.[435a] *Melobesia patena* Hooker fils et Harvey in Harvey.[199]

Main Characteristics:
—Hypothallus coaxial
—Cell fusions restricted to conceptacle roofs
—Epithallial cells rounded
—Spermatangia produced in dendroid branching filaments

Main Reference: Townsend.[435a]

Tenarea Bory.[489] Type Species: *T. undulosa* Bory.[489]

Main Characteristics:
—Hypothallus a single layer of elongated, oblique cells
—Habit of erect plates anastomosing, coalescing, and abutting

Main References: H. Huvé,[215] Adey,[13] and Cabioch.[76]

Comments: See the comments under *Dermatolithon*.

Yamadaea Segawa.[389] Type Species: *Y. melobesioides* Segawa.[389]

Main Characteristics:
—Conceptacles axial in origin
—Fronds consisting of one to two intergenicula only
—Crustose base extensive

Main References: Dawson and Steele;[125] Abbott and Hollenberg.[1]

Comments: *Yamadaea americana* was described from the San Juan Islands, Washington, but it may not be distinct from the type species from Japan. *Yamadaea melobesioides* has been reported from Monterey, Calif. Segawa's[389] careful description revealed that this genus is like *Corallina* in all respects except frond size.

APPENDIX 2

SPECIES IN THE CORALLINACEAE

The list below contains those species mentioned in the text as well as others that are relatively common. The list is by no means complete.

Alatocladia modesta (Yendo) Johansen
Amphiroa beauvoisii Lamouroux
A. crassa Lamouroux
A. crustiformis Dawson
A. dilatata Lamouroux
A. echigoensis Yendo
A. ephedraea (Lamarck) Decaisne
A. foliacea Lamouroux
A. fragilissima (Linnaeus) Lamouroux
A. hancockii Taylor
A. misakiensis Yendo
A. pusilla Yendo
A. rigida Lamouroux
A. rigida var. *antillana* BØrgesen
A. tribulus (Ellis et Solander) Lamouroux
A. valonoioides Yendo
A. zonata Yendo
Antarcticophyllum aequabile (Foslie) Mendoza
A. subantarcticum (Foslie) Mendoza
Arthrocardia corymbosa (Lamarck) Decaisne
A. duthiae Johansen
A. gardneri Manza
A. papenfussii Manza
A. silvae Johansen
A. stephensonii Manza
Bossiella californica (Decaisne) Silva
B. californica ssp. *schmittii* (Manza) Johansen
B. chiloensis (Decaisne) Johansen
B. cretacea (Postels et Ruprecht) Johansen
B. orbigniana (Decaisne) Silva
B. orbigniana ssp. *dichotoma* (Manza) Johansen
B. plumosa (Manza) Silva
Calliarthron cheilosporioides Manza
C. latissimum (Yendo) Manza
C. tuberculosum (Postels et Ruprecht) Dawson
C. yessoense (Yendo) Manza
Chaetolithon deformans (Solms-Laubach) Foslie
Cheilosporum cultratum (Harvey) J. Areschoug
C. jungermannioides Ruprecht in J. Areschoug
C. proliferum (Lamouroux) DeToni
C. sagittatum (Lamouroux) J. Areschoug
Chiharaea bodegensis Johansen
Choreonema thuretii (Bornet) Schmitz
Clathromorphum circumscriptum (Strömfelt) Foslie
C. compactum (Kjellman) Foslie
C. loculosum (Kjellman) Foslie
C. nereostratum Lebednik
C. parcum (Setchell et Foslie) Adey
C. reclinatum (Foslie) Adey
Corallina confusa Yendo
C. cubensis (Montagne) Kützing
C. elongata Ellis et Solander
C. kaifuensis Yendo

C. officinalis Linnaeus
C. pilulifera Postels et Ruprecht
C. sessilis Yendo
Dermatolithon corallinae (Crouan) Foslie
D. cystoseirae (Hauck) H. Huve
D. pustulatum (Lamouroux) Foslie
D. stephensoni Lemoine
Ezo epiyessoense Adey, Masaki et Akioka
Fosliella farinosa (Lamouroux) Howe
F. farinosa f. *solmsiana* (Falkenberg) Foslie
F. lejolisii (Rosanoff) Howe
F. limitata (Foslie) Ganesan
F. minutula (Foslie) Ganesan
F. minutula f. *lacunosa* Suneson
F. *paschalis* (Lemoine) Setchell et Gardner
F. tenuis Adey et Adey
F. valida Adey et Adey
F. zonalis (Crouna et Crouan) Ganesan
Goniolithon byssoides (Lamarck) Foslie
G. papillosum (Zanardini ex Hauck) Foslie
Haliptilon cuvieri (Lamouroux) Johansen et Silva
H. gracilis (Lamouroux) Johansen
H. squamatum (Linnaeus) Johansen, Irvine et Webster
H. subulatum (Ellis et Solander) Johansen
Heteroderma nicholsii Setchell et Mason
H. subtilissima (Foslie) Foslie
Hydrolithon borgenseni (Foslie) Foslie
H. decipiens (Foslie) Adey
H. reinboldii (Weber-van Bosse et Foslie) Foslie
Jania adherens Lamouroux
J. capillacea Harvey
J. corniculata (Linnaeus) Lamouroux
J. crassa Lamouroux
J. longifurca Zanardini
J. nipponica (Yendo) Yendo
J. pumila Lamouroux
J. radiata (Yendo) Yendo
J. rubens (Linnaeus) Lamouroux
J. verrucosa Lamouroux
Kvaleya epilaeve Adey et Sperapani
Leptophytum foecundum (Kjellman) Adey
L. laeve (Strömfelt) Adey
Litholepis caspica (Foslie) Foslie
L. expansum Philippi
L. mediterranea Foslie
Lithophyllum daedaleum Foslie et Howe
L. incrustans Philippi
L. kotschyanum Unger
L. moluccense Foslie
L. nitorum Adey et Adey
L. orbiculatum (Foslie) Lemoine
L. tortuosum Foslie
Lithoporella melobesioides (Foslie) Foslie

L. pacifica (Heydrich) Foslie
Lithothamnium australe Foslie
L. coralloides Crouan et Crouan
L. dickiei Foslie
L. glaciale Kjellman
L. indicum Foslie
L. lemoineae Adey
L. ruptile (Foslie) Foslie
L. sonderi Hauck
L. soriforum Kjellman
L. tophiforme Unger
L. ungeri Kjellman
Lithothrix aspergillum Gray, J. E.
Marginisporum aberrans (Yendo) Johansen et Chihara
M. crassissimum (Yendo) Ganesan
M. declinata (Yendo) Ganesan
Mastophora licheniformis Decaisne
M. macrocarpa Montagne
Mastophoropsis canaliculata (Harvey in Hooker) Woelkerling
Melobesia marginata Setchell et Foslie
M. mediocris (Foslie) Setchell et Mason
M. membranacea (Esper) Lamouroux
M. vanheurkii (Heydrich) Cabioch
Mesophyllum aleuticum Lebednik ined.
M. conchatum (Setchell et Foslie) Adey
M. lichenoides (Ellis et Solander) Lemoine
Metagoniolithon chara (Lamarck) Ducker
M. chara var. *dichotomum* Ducker
M. radiatum (Lamarck) Ducker
M. stelliferum (Lamarck) Weber-van Bosse
Metamastophora flabellata (Sonder) Setchell
Neogoniolithon absimile (Foslie et Howe) Cabioch
N. accretum Adey et Vassar
N. fosliei (Heydrich) Setchell et Mason,
N. frutescens (Foslie) Setchell et Mason
N. megacarpum Adey
N. myriocarpum (Foslie) Setchell et Mason
N. notarisii (Dufour) Hamel et Lemoine
N. spectabile (Foslie) Setchell et Mason
N. strictum (Foslie) Setchell et Mason
N. westindianum Adey
Phymatolithon calcareum (Pallas) Adey et McKibbin
P. laevigatum (Foslie) Foslie
P. lenormandii (Areschoug) Adey
P. polymorphum (Linnaeus) Foslie
P. rugulosum Adey
Porolithon craspedium (Foslie) Foslie
P. gardineri (Foslie) Foslie
P. onkodes (Heydrich) Foslie
P. pachydermum (Weber-van Bosse et Foslie) Foslie
Schmitziella cladophorae Chapman
S. endophloea Bornet et Batters in Batters
Serraticardia macmillanii (Yendo) Silva
S. maxima (Yendo) Silva
Sporolithon erythraeum (Rothpletz) Kylin
S. ptychoides Heydrich
Synarthrophyton patena (Hooker fils et Harvey) Townsend
Tenarea confinis (Crouan et Crouan) Adey et Adey
T. hapalidioides (Crouan et Crouan) Adey et Adey
T. prototypa (Foslie) Adey
T. tessellatum (Lemoine) Littler
T. undulosa Bory
Yamadaea americana Dawson et Steele
Y. melobesioides Segawa

INDEX

A

Accretion rates, see also Growth rates, 149
Acetabularia, calcifying algae in, 111
Acrochaetium, biotic interactions of, 146
Acropora, 168
  palmata, 169
Adaptations, of coralline algae, 188—191
Alaska, and phytogeography of coralline algae, 129
Alaskan earthquake, and uplifted coralline shores, 140
Alatocladia
  branching in, 66
  classification of, 215
  conceptacles of, 83
  phytogeography of, 123, 130
Algae, see also Coralline algae
  endolithic, 161
  noncalcareous, 152
Algal cup reefs, 166, 167
Algal ridges, 159—160
  formed from fused cup reefs, 167
  in Caribbean, 164—166
  Waikiki reef, 163
Alizarin Red-S, 153
Amphiroa
  branching, 74
  classification of, 215
  conceptacles, 82
  effect of light on, 138
  effect of seasons on, 137
  epithallia of, 30
  female conceptables in, 100
  genicula, 68—72
  magnesium content in, 112
  male conceptacles in, 93
  "minisegments" in, 72
  phytogeography of, 123, 127, 129—132
  secondary pit-connections in, 16
  spore germination in, 26
  sporeling development in, 24
  taxonomic systems for, 179
Amphiroa anceps, 3
  branching in, 68
  genicula in, 70
  secondary pit-connections in, 18
  water motion and, 140
Amphiroa beauvoisii, genicula in, 70
Amphiroa bowerbankii, genicula in, 70, 72
Amphiroa capensis, genicula in, 70, 72
Amphiroa ephedraea
  branching in, 68
  fusion cells in, 104
  genicula in, 69
  gonimoblast filaments in, 105
  male conceptacles in, 94
  spores and substrates, 135
  tetrasporangial conceptacles in, 88

Amphiroa fragilissima
  calcification rate for, 114
  genicula in, 72
Amphiroa rigida, cell walls of, 13
Amphiroa subcylindrica, branching in, 68
Amphiroa valonioides, branching in, 68
Amphiroa zonata, growth forms in, 5
"*Amphiroa* zone", 132
Amphiroeae (subfamily), taxonomic system for, 189
Amphiroideae (subfamily), 10, 66—67
  Amphiroa, 67—68
  genicula of, 68—72
  Lithothrix, 72—73
  meristematic cells of, 27
  reproductive initials in, 91
  secondary pit-connections in, 16, 20
  spore germination in, 24
Antarctic, see also Phytogeography, 120
Antharcticophyllum
  classification of, 215
  thick crusts of, 47
Appearance, of coralline algae, 1
Aragonite, in biominialization, 111, 156
Archaeolithophyllum, 176
Archaeolithothamnium
  epithallia of, 31
  taxonomic system for, 182
Arctic, see also Phytogeography and effect of temperature, 137—138
Argentina, and phytogeography of coralline algae, 133
Arthrocardia
  branching in, 65, 66
  classification of, 215—216
  conceptacles of, 83
  intergenicula of, 58
  male conceptacles in, 93
  phytogeography of, 123, 130
Atlantic Ocean
  northeastern, 124—127
  northwestern, 119—124
  temperature related distributions, 137—138
Atoll(s), see also Reef
  algal cup reefs, 166
  Pacific, 159
Attachment discs, 79
Australia, and phytogeography of coralline algae, 120, 133, 138
Autolysis, in tetrasporangial conceptacle development, 85
Autotrophy, 1
Auxiliary cells
  and gonimoblast filaments, 105
  in carpogonia, 99

B

Barbados, Northern Bellairs Reef of, 153

Basal cells, in male conceptacles, 97
Bermuda
   algal cup reefs of, 166, 167
   coralline algae of, 141
Bicarbonate usage theory, 115
Biogeography, see Phytogeography
Bioherms
   characteristics of, 159
   of St. Croix reef system, 168
Biomineralization, see also Calcification
   physiology of, 9
   substances involved in, 156
Biospores, 22, 23
Biotic interactions, 143—147
Bisporangia
   development of, 91—92
   structure of, 7
Bisporangial plants, 108
Bisporangial producer, facultative, 108
Boreal group, effect of temperature on, 137—138
Boring animals, and ridge formation, 169
Boring macroorganisms, cavities of, 161
Bossiella
   and Alaskan earthquake, 140
   biotic interactions of, 146
   branching in, 66
   calcification in, 14, 113
   calcium carbonate accretion, 154
   classification of, 216
   conceptacles in, 80, 83
   epithallia of, 30
   female conceptacles in, 100
   intergenicula of, 57
   phytogeography of, 123, 130, 131, 133
   sewage, 143
   tetrasporangial conceptacles in, 85
Bossiella californica spp. *Schmitti*
   growth forms in, 5
   intergenicula of, 58
Bossiella cretacea, meristematic cells in, 27
Bossiella orbigniana
   bisporangial plants of, 108
   branching in, 63
   growth spurts in, 28
Bossiella plumosa, growth forms in, 5
Branch(es)
   in *Lithothrix*, 72
   primary vs. secondary, 63
   system of, 57
Branching
   Amphiroa, 67
   corallinoideae, 63—65
   dendroid, 95
   dichotomous, 64, 75
   in taxonomic system, 186
   pinnate, 64
Browsing animals, see also Grazing, 145, 149
Brucite, in cell walls, 112

C

Calcification
   cell wall composition, 111—112
   mechanisms for, 116
   process of, 13—15, 112—117
Calcification rates
   determination of, 113—114
   variation in, 152—153
Calcite
   fate of, 154—156
   in biomineralization, 111
Calcite crystals, growth of, 15—16
Calcium carbonate
   accretion in coral reefs, 154
   in cell walls, 1, 13
   in tropical bioherm environment, 155
   in tropical waters, 137
   production of, 152—153
California
   and calcium carbonate accretion, 154
   and phytogeography of coralline algae, 128, 129, 131, 132
   sewage studies in, 143
Calliarthron, 59
   and calcium carbonate accretion, 154
   biotic interactions of, 146
   branching in, 66
   calcification in, 14
   classification of, 216
   conceptacles of, 83
   effect of seasons on, 137
   nuclei of, 20
   phytogeography of, 123, 130, 131
   tetrasporangial conceptacles in, 85
Calliarthron cheilosporioides
   fronds of, 58
   growth forms in, 5
   growth rates for, 150
   succession in, 150
Calliarthron tuberculosum
   branching in, 63
   branch tip of, 150
   carpogonia development in, 102
   carpogonia in, 101
   conceptacles in, 81, 84
   female conceptacles in, 100
   frond imitation in, 79
   frond initiation in, 65, 66
   fusion cells in, 104
   genicula in, 60
   gonimoblast filaments in, 105
   host for *Clathromorphum parcum*, 49, 51
   intergenicula of, 58
   tetrasporangial conceptacles in, 87
Calorific values, for benthic algae, 145
Carbon dioxide utilization theory, of carbonate deposition, 114
Carbonia anhydrase, in calcification process, 114
Caribbean Sea
   algal ridge areas of, 165—166
   coralline ridge-formers in, 168—170
   Pleistocene platforms of, 159
   reefs, 164—168
Carpogonia, 98—101

Carpogonial filaments, 100, 101
Carpogonial initials, 98
Carpospores, 23, 24
Carposporophytes, 7, 101—106
Cations, in coralline algae, 111
Caulerpales, calcifying algae in, 111
Cell fusions, see also Fusion cells
   photomicrograph of, 33
   stages in development of, 19
Cellulose
   in calcification process, 114
   in cell walls, 13
Cell walls
   calcium carbonate in, 1
   composition of, 111—112
   structure of, 13
Cementation
   and extracellular products, 156
   internal carbonate, 155
Chaetolithon
   classification of, 216
   parasitic existence of, 49
Cheilosporum
   branching in, 66
   classification of, 216
   conceptacles of, 83
   female conceptacles in, 100
   host to *C. thuretti*, 53
   male conceptacles in, 94
   phytogeography of, 123, 130
   tetrasporangial conceptacles in, 87
Cheilosporum cultratum
   conceptacles in, 80
   intergenicula in, 59
Cheilosporum proliferum, 3, 63
   branching in, 63
   growth spurts in, 28
Cheilosporum sagittatum, growth forms in, 5
Chemical erosion, 155
Chiharaea
   adaptation of, 190
   branching in, 66
   classification of, 216—217
   conceptacles of, 83
   intergenicula of, 57
   phytogeography of, 123, 130
Chiharaea bodegensis
   growth forms in, 5
   intergenicula of, 58
Choreonema
   classification of, 217
   parasitic existence of, 49
   parasitic species of, 53
   phytogeography of, 125, 128
   spore germination in, 26
Choreonema thuretti, parasitic existence of, 53, 54
Chromosome complements, determination of, 20
Chromosome numbers, in Corallinaceae, 37
Cladophora feredayi, parasitic existence of, 55
Cladophora pellucida, parasitic existence of, 55

Cladostephus, biotic interactions of, 146
Clathromorphum
   adaptation of, 190
   carposporophytes in, 105
   cellular connections in, 16
   classification of, 217
   conceptacles in, 81
   crust morphology of, 48
   cuticle on surface of, 31
   epiphytic species in, 49, 52
   epithallial cells in, 29, 31
   fossil, 173
   growth spurts in, 28
   hypothallus in, 40
   male conceptacles in, 95, 96—97
   phytogeography of, 119, 120, 122, 124, 125, 128, 129
   tetrasporangial conceptacles in, 89
   thick crusts of, 47, 48
   ventral perithallus in, 40
Clathromorphum circumscriptum
   and Alaskan earthquake, 140
   biotic interactions of, 144—145
   calcification process in, 115
   conceptacle initiation, 136
   hypothallus in, 40
   marginal growth rates of, 137
Clathromorphum compactum, hypothallus in, 40
Clathromorphum nereostratum
   conceptacles in, 81
   crust morphology of, 48
   tetrasporangia in, 90
Clathromophum parcum
   diagnostic features of, 49—50
   epiphytic existence of, 49, 51
   vertical section through, 51
Clathromorphum reclinatum
   and Alaskan earthquake, 140
   conceptacles, 82
   epiphytic existence of, 49, 50
Coccolithophorids, calcite-producing, 111
Colliarthron tuberculosum, growth rate for, 149
Colonization, 150—151
Columella, 89
Conceptacles, 79—84
   axial, 82
   Corallinoideae, 83
   cortical, 82
   diagrams of, 8
   evolution of, 81
   female, 100
   impact of seasonality on, 136
   intergenicula, 58
   male, 92—97
      development of, 95
      in *Clathromorphum*, 96—97
      in taxonomic system, 185, 187
   marginal, 82
   mature, 84, 97
   multi-pored, 80
   pseudolateral, 83

shape of, 93
single-pored, 80
tetrasporangial, 84—91
   immature, 87
   one-pored, 41
   types of, 84—91
types of, 6—7, 82
Conceptacular cap, 84
Connecting filaments, 106
Coral animals, calcification in, 114
Corallina, 10
   and Alaskan earthquake, 140
   branching in, 66
   calcification in, 13—16
   calcium carbonate accretion, 154
   chromosome number, 20
   classification of, 217
   conceptacles of, 83
   effect of seasons on, 137
   epithallial cells of, 30
   genicula in, 61
   intergenicula of, 57
   male conceptacles in, 93
   phytogeography of, 123, 125, 126, 130, 131
   taxonomic systems for, 179
   tetrasporangial conceptacles in, 85
Corallinaceae (family), 1, 173
   articulated, 12
   chromosome numbers in, 37
   fossil, 176—177
   geological ages of, 175
   nonarticulated, 11
   species in, 225—226
Corallina cuvieri, epithallial cells in, 30
Corallina elongata
   frond initiation in, 65
   meiosis in, 92
Corallina officinalis
   absence of epithallial cells in, 32
   biotic interactions of, 144
   branching in, 28, 63
   calcification process in, 115
   calcification rates in, 113
   cell fusion in, 19
   effect of light on, 138
   effect of seasons on, 136
   epithallial cells in, 29, 30
   frond initiation in, 65
   geniculum formation in, 27—28
   Golgi bodies in, 21
   growth forms in, 5
   growth rates for, 150
   host to *C. thuretti*, 53
   intergenicula of, 58
   in vermifuge preparations, 147
   life history for, 106—107
   male conceptacles in, 94, 97
   meiosis in, 92
   phytogeography of, 124
   plastid development in, 21
   primary pit-connections in, 16, 17
   pseudodichotomous branching in, 28
   sewage, 143
   spore germination in, 22—24
   spores and substrates, 135—136
   taxonomic systems for, 179
   var. *chilensis*, 151, 152
   water motion, 140
Corallina pilulifera
   and desiccation, 139—140
   effect of light on, 138
   succession in, 150
Corallineae (tribe), 10, 182, 189
Coralline algae
   adaptations in, 188—191
   Amphiroideae, 66—67
   articulated, 2—3
      Amphiroa, 67—68
      branching of, 63—65
      cell fusions, 19
      cellular connections of, 16—20
      classification of, 10, 11
      frond initiation, 65—66
      genicula, 60—62, 68—72
      growth, 5
      in fossilized state, 177
      intergenicula of, 57—60
      Lithothrix, 72
      Metagoniolithoideae, 74—76
      phytogeography of, 123, 130, 131
      taxonomy and nomenclature of, 181, 184, 188
   basic structure of, 1—7
   cellular structure of, 20—22, 37
   cell walls of, 13—16
   classification of, 10
   epithallia, 28—32
   life cycle in, 9
   meristems, 26—28
   nonarticulated, 1—2
      classification of, 10, 11
      Dermatolithon, 42—44
      epiphytic, 48—52
      Foslie classification scheme for, 183
      hosts for, 49
      lithophylloideae, 41
      Lithophyllum series, 42
      Mastophoroideae, 44
      Melobesioideae, 44
      parasitic, 52—55
      phytogeography of, 122, 128
      ribbon corallines, 46
      taxonomy and nomenclature of, 181, 188
      thick crusts, 47—48
      thin crusts, 44—46
      unattached, 48
   reef-building, 162
   role in oceans, 8—9
   spore generation, 22
   sporeling growth and development, 24—26
   trichocytes and megacells, 32—35
   unattached, 153—154

Corallinoideae (subfamily), 10
  branching of, 63
  conceptacles of, 83
  fossil, 173
  frond initiation in, 65—66
  genicula of, 60—62
  intergenicula of, 57—60
  reproductive initials in, 91
  sporeling development in, 24
  tabular key to, 186—187
  taxonomic system for, 186—187
Corals
  productivity of, 152
  taxonomic systems for, 179
Cortex, structure of, 4
Cortical filaments, 59
Cretacicrusta dubiosia, 174
Crusts, see also Epithallia, Hypothallus
  composition of, 40
  hypothallial and perithallial tissue in, 48
  in Lithophylloideae, 43
  thick, 47—48
Cryptocrystallization, 155, 161
Cryptonemiales (order), 1, 12, 111
Crystallization
  aragonite, 156
  nonbiological, 155
Crystals, 37
Curacao
  and phytogeography of coralline algae, 130
  Porolithon-coral ecosystem of, 161
  reef building on shores of, 152, 164, 169
Cuticle(s)
  characteristics of, 31
  in conceptacle development, 86
  laminated, 32
  of epithallia, 29
Cymopolia, calcifying algae in, 111
Cystophora, epiphytism of, 146
Cytokineses, 92
Cytology
  cell fusions, 19—20
  cell walls, 13—16
  epithallia, 29—32
  meristems, 26—28
  nuclei, see also Nuclei, 20
  organelles and inclusions, 20—22, 37
  primary pit-connections, 16, 23, 58
  secondary pit-connections, 16—18, 41
  spore germination, 22—24
  sporeling growth and development, 24—26
  trichocytes and megacells, 32—35

D

Dasycladales, calcifying algae in, 111
Day length, effect on growth rates, 137
Decalcification, 16, 70
Depth, and water motion, 141
Deposition, calcification process and, 117

Dermatolithon
  cell fusions in, 19
  chromosome number of, 20
  classification of, 217
  crust type of, 43
  diagnostic features of, 42
  effect of seasons on, 137
  epiphytic existence of, 49
  life history for, 108
  phytogeography of, 125, 128
  secondary pit-connections in, 16
  starch grains in, 21—22
Dermatolithon corallinae, life history for, 108
Desiccation, see also Environment
  and growth rates, 139—140
  effect on coralline aglae, 8
Development, post-fertilization, 103
Dioeciousness, 84
"Discoid bodies", 79
Distributional studies, see also Phytogeography, 119
Dolomite, 156
Dumontia, spore germination in, 22—23

E

Echinometra, 161
Endoplasmic reticulum
  and starch grain formation, 22
  nuclear, 92
Eniwetok atoll, reef growth on, 152—153
Environment, see also Light, Seasons, Water motion
  and growth, 65
  intergenicular morphology and, 57
  of coralline algae, 8
  seasonal effects, 107, 136
  trichocyte development, 34
Epiphytes, fouling, 189
Epithallia
  characteristics of, 29
  function of, 30
  in taxonomic system, 185
  multistratose, 31
  structure of, 4, 31
Erosion, chemical, 155
Ezo
  classification of, 217
  diagnostic features of, 42
  parasitic existence of, 49
  parasitic species of, 52
Ezo epiyessoense, parasitic existence of, 52

F

"Faux hypothalle"
  in *Goniolithon*, 43
  in *Lithophyllum*, 42
Filamentous red algae, biotic interactions of, 146
Filaments

characteristics of, 39
of parasitic species, 55
perithallial, 40
Florideophyceae (class), 12
Foslie classification scheme, 183
Foslie, M., 182
Fosliella
cell walls of, 13
chromosome number of, 20
classification of, 217—218
epithallial cells of, 30
phytogeography of, 125, 128
sporeling development in, 24, 25
surface view of, 45
trichocyte development in, 32, 34
with thin crust, 44
Fosliella farinosa
meiosis in, 92
trichocytes in, 33
Fosliella farinosa f. *solmisiana*, diagnostic features of, 44—45
Fosliella lejolisii, with thin crust, 44
Fosliella limitata, with thin crust, 44
Fosliella paschalis, dispersal unit produced by, 80
Fosliella zonalis, cell walls in, 15
Fossil coralline algae
ancestral, 175—176
appearance of corallinaceae, 176—177
extant genera reported as, 177
Solenoporaceae, 173—175
Fouling epiphytes, 189
Frond, system of, 57
Frond initiation, in corallinoideae, 65—66
Fungi
ascomycetous, 145
endolithic, 161
Fusion cells, 103, 104
formation of, 106
in taxonomic system, 187
Fusions, 16

G

Galaxaura, aragonite-producing, 111
Genicula
in *Amphiroa*, 68—72
in Corallinoideae, 60—62
in *Metagoniolithon*, 74—75
structure of, 2, 6
Genicular cells, 27, 61
Geography, see Phytogeography
Golgi bodies, 21, 37
Gonimoblast filaments, 103, 104
Goniolithon
brucite in, 112
classification of, 217
crust types of, 43
diagnostic features of, 43—44
phytogeography of, 122
Goniolithon byssoides, diagnostic features of, 44

Goniolithon papillosum, diagnostic features of, 44
Grazing
and reef production, 164
positive effects of, 161
Grazing animals, and uncalcified cell parts, 29
Growth
characteristics of, 8
colonization and succession, 150—151
effect of light on, 138—139
effect of water motion, 140—141
genicular, 62, 71
marginal, 47—48
seasonality and, 136—137
seawater ingredients, 142
Growth lines, in perithallial tissues, 28
Growth rates
and desiccation, 139—140
and sewage, 143
effect of light on, 137
effect of temperature on, 137—138
for reef communities, 152—153
studies of, 149—150
Gut contents, examinations of, 145
Gymnocodiaceae (family), 173

H

Habitat, in taxonomic system, see also Environment, 184, 185
Halimeda, calcifying algae in, 111
Haliptilon
branching in, 65, 66
classification of, 218
conceptacles of, 83
host of *C. thuretti*, 53, 54
phytogeography of, 123, 125, 129—131
sewage, 143
Haliptilon cuvieri:
epithallial cells of, 30
primary pit-connections in, 16, 17
water motion, 140
Haliptilon squamatum
frond initiation in, 65
host to *C. thuretti*, 54
gonimoblast filaments in, 105
phytogeography of, 126
Hawaiian Islands
and coralline productivity, 151
and phytogeography of coralline algae, 138
reefs of, 160, 162
Heteroderma
classification of, 218—219
epiphytism of, 146
phytogeography of, 128
trichocytes in, 35
with thin crust, 44
Heydrich, F., 182
Hokkaido
and phytogeography of coralline algae, 127, 129

and succession in coralline algae, 150
Holdfasts, crustose, 76
Host, coralline algae as, 49
Hydrogen ions, in calcification process, 117
Hydrolithon
   classification of, 219
   effect of light on, 138
   phytogeography of, 122, 127—129
   reef building of, 160—162
   trichocytes in, 35
Hydrolithon reinboldii, habitat of, 162
Hypothallus
   characteristics of, 39
   Clathromorphum parcum, 50
   coaxial, 39, 40, 42
   formation of, 41
   in taxonomic system, 185
   primary, 40
   secondary, 28, 40, 42
   structure of, 3, 6

I

Iceland, and phytogeography of coralline algae, 140
Indian Ocean, 120
   and phytogeography of coralline algae, 132
   reef production in, 162—163
Intercalary meristems, 27
Intergenicula
   branches arising from, 66
   in Corallinoideae, 57
   in *Lithothrix*, 72
   in taxonomic system, 187
   structure of, 2, 6

J

Jania
   absence of epithallial cells in, 32
   branching in, 65, 66
   chromosome number of, 20
   classification of, 219
   conceptacles of, 83
   epithallia of, 30
   host of *C. thuretti*, 53
   male conceptacles in, 94
   phytogeography of, 123, 125, 216, 129—132
   taxonomic systems for, 179
   tetrasporangial conceptacles in, 87
Jania capillacea
   deciduous propagule of, 80
   growth forms in, 5
   intergenicula of, 57
Jania coniculata, trichocytes in, 33
Jania fastigiata, and water motion, 140
Jania prolifera, branching in, 63
Jania pusilla, epiphytism of, 146
Jania rubens
   epithallial cells of, 29
   frond initiation in, 65, 79
   intergenicula in, 59
   intergenicular and genicular cells in, 62
   life history for, 107
   male conceptacles in, 97
   meiosis in, 92
   meristematic cells in, 26
   plastid development in, 21
   spore germination in, 22—24
   tetraspore development of, 135—136
   trichocyte development in, 32—34
   vermifuge preparations, 147
Janieae, taxonomic system for, 189
Japan
   and phytogeography of coralline algae, 127—129
   succession in coralline algae of, 150
Joculator, conceptacles of, 84
Jurassic, appearance of Corallinaceae during, 176

K

Keega, 173
Kelp beds, and coralline algae colonization, 150, 151
Kvaleya
   classification of, 219
   gonimoblast filaments in, 106
   parasitic existence of, 49, 52
   phytogeography of, 125
Kvaleya epilaeva, parasitic existence of, 52

L

Lava flows, and coralline algae colonization, 150
Leptophytum
   cellular connections in, 16
   classification of, 219—220
   conceptacles, 82, 84
   epithallia of, 31
   gonimoblast filaments in, 106
   phytogeography of, 122, 125, 126, 128
   thick crusts of, 47
Leptophytum laeve, as host to *K. epilaeve*, 52
Liagora, aragonite-producing, 111
Life cycle, in coralline algae, 9
Life histories, 7, 106—108
Light, see also Environment, 8
   and calcification rate, 114
   and reef production, 164
   effect on growth rates, 137—139
Limpets, 144
Litholepis
   classification of, 220
   diagnostic feature of, 45
Lithophylloideae (subfamily), 10
   diagnostic features of, 41
   meristematic cells of, 27

ribbon growth form in, 46
secondary pit-connections in, 16, 20
sporeling development in, 24
spore germination in, 22
Lythophyllum, 2
and calcium carbonate accretion, 154
cell fusions in, 19
classification of, 220
diagnostic features of, 42
effect of seasons on, 136
epithallial cells of, 30
fossil, 173
growth spurts in, 28
hypothallus in, 40
life history for, 108
magnesium content in, 112
Mediterranean reefs of, 155
phytogeography of, 122, 126, 128, 129
productivity of, 151, 152
reef building of, 160—162
starch grains in, 21—22
unattached, 48
Lithophyllum congestum
growth rates for, 149
in ridge formation, 168, 169
water motion, 140
Lithophyllum expansum, secondary pit-connections in, 18
Lithophyllum incrustans
cell walls of, 13
epithallial cells of, 29, 31
hypothallus in, 42
Lithophyllum kotschyanum, reef production of, 164
Lithophyllum orbiculatum
hypothallus in, 32
secondary pit-connections in, 18
Lithophyllum tortuosum, secondary pit-connections in, 18
Lithophyllum yessoense, as *Ezo epiyessoense* host, 42, 52
Lithoporella
classification of, 220
diagnostic features of, 46
phytogeography of, 122, 128
Lithoporella melobesioides, conceptacles in, 80
Lithothamnium
and Alaskan earthquake, 140
and calcium carbonate accretion, 154
cellular connections in, 16
classification of, 220—221
effect of seasons on, 137
epithallial cells in, 29, 31
fossil, 173
gonimoblast filaments in, 106
magnesium content in, 112
meristematic cells in, 26
phytogeography of, 119, 121, 122, 124—129
taxonomic system for, 182
thick crusts of, 47
unattached, 48

Lithothamnion ball, 154
Lithothamnium corallioides
in commercially collected marl, 146
water motion, 141
Lithothamnium glaciale, effect of light on, 138
Lithothamnium pacificum
conceptacles in, 80
male conceptacles of, 96
Lithothamnium parcum, epiphytic existence of, 49
Lithothamnium sonderi, cellular structures in, 22
Lithothrix
branching in, 74, 79
classification of, 221
phytogeography of, 123, 130, 131
Lithothrix aspergillum
biosporangial plants of, 108
branching in, 73
gonimoblast filaments of, 103
meristematic cells in, 26, 27
sewage, 143
spermatangia in, 98
sporeling development in, 24
starch grain formation, 22
Lithotricheae, taxonomic system for, 189

M

Macrocystis bed, 150
Magnesian calcite, 156
Magnesium
and calcite production, 156
and water temperature, 138
in cell walls, 111
Maine, Gulf of, and phytogeography of coralline algae, 138
Marginal growth, 47—48
Marginisporum
branching in, 66
chromosome number of, 20
classification of, 221
conceptacles of, 83
effect of seasons on, 137
phytogeography of, 123, 130
Marl
anastomosing of protuberances of, 142
chracteristics of, 154
commercial uses of, 146—147
defined, 2
phytogeography of, 131—132
samples of, 47
soil condition, 146
structural features of, 48
water motion, 141
Mastophora
classification of, 221
diagnostic features of, 46
phytogeography of, 128
Mastophora macrocarpa, structural features of, 46

Mastophoroideae (subfamily), 10, 44
   ribbon growth form in, 46
   sporeling development in, 24
   taxonomic system for, 185—186
Mastophoropsis
   classification of, 221
   hypothallus in, 40
   ribbon growth form in, 46
   tetrasporangial conceptacles in, 89
Mastophoropsis canaliculata, ribbon-like thallus of, 45
Mediterranean Sea, and phytogeography of coralline algae, 120, 133
Medulla, structure of, 4
Medullae, *Amphiroa*, 68
Medullary cells, 59
   C. tuberculosum, 149
   derived from terminal meristems, 27
   in *Metagoniolithon*, 76
Megacells, 79
   as diagnostic features, 35
   development of, 32—34
   in *Porolithon*, 35
   in taxonomic system, 186
Meiosis, 92, 107
Melobesia
   carposporophytes in, 105
   classification of, 221
   phytogeography of, 125, 128
   taxonomic systems for, 179
   with thin crust, 44
Melobesia membranacea, with thin crust, 44
Melobesia pacifica
   biotic interactions of, 144
   sporeling development in, 24, 25
Melobesia van heurckii, lack of calcium carbonate in, 16
Melobesieae (tribe), 10, 182
Melobesioideae (subfamily), 6, 10, 44
   cell fusions in, 20
   cellular structures in, 22
   conceptacle formation in, 91
   fossil, 173
   fusion cells in, 106
   male conceptacles in, 93
   parasitic species of, 52
   phytogeography of, 119—120
   reproductive initials in, 91
   ribbon growth form in, 46
   secondary pit-connections in, 16
   sporeling development in, 24
   tabular key to, 184—185
   thick crusts of, 47
Meristems
   characteristics of, 26
   cortical filaments, 59
   in *Fosliella*, 24
   in *Metagoniolithon*, 74
   intercalary, 4, 7
   primary, 4, 7
   pseudodichotomous branching in, 28
   types of, 27

Mesophyllum
   carposporophytes in, 105
   cell divisions in, 16
   classification of, 221
   crust morphology of, 48
   epiphytic existence of, 49, 52
   hypothallus in, 39, 40
   male conceptacles in 95—96
   phytogeography of, 119, 122, 125, 126, 128, 129
   tetrasporangial conceptacles in, 89
   thick crusts of, 47
   ventral perithallus in, 40
Mesophyllum conchatum
   biotic interactions of, 145
   carposporophyte of, 107
   diagnostic features of, 51
   epiphytic existence of, 49, 50
   tetrasporangia in, 90
Mesophyllum lamellatum, epiphytic existence of, 52
Mesophyllum lichenoides
   cell fusion in, 19
   epithallial cells in, 29, 31
Metagoniolithoideae (subfamily), 10, 74—76
Metagoniolithon
   cell fusions in, 19
   classification of, 222
   epiphytism of, 146
   epithallia of, 30
   genicula in, 74—76
   meristematic cells in, 27
   phytogeography of, 123, 130, 133
Metagoniolithon charoides, and water motion, 140
Metagoniolithon radiatum, 3
   growth forms in, 5
   trichocyte development in, 32, 33, 34
Metagoniolithon stelliferum, 75, 145
Metamastophora
   chromosome number of, 20
   classification of, 222
   hypothallus in, 40
   male conceptacles in, 97
   phytogeography of, 122
   ribbon growth form in, 46
Metamastophora flabellata
   diagnostic features of, 44
   hypothallus in, 40
   structural features of, 46
Microatolls, 166
Milrow Nuclear test, and uplifted coralline shores, 140
"Minisegments", 72
Mitosis, 20
Monoeciousness, 84
Mother cells, spermatangial, 93, 96, 97, 99
Movement, see Water motion
Multiporate development, 84
Mussels, and coralline algae succession, 151
Nycophycophila, biotic interactions of, 145

N

Nacarria, spore germination in, 22, 26
Nemaliales, aragonite-producing, 111
Neogoniolithon, 2
　classification of, 222
　fusion in, 18
　in ridge formation, 168, 169
　magnesium content in, 112
　megacell development in, 34
　phytogeography of, 119, 122, 128, 129
　productivity of, 151, 152
　protuberance formation in, 48
　reef building of, 160—161
　thick crusts of, 47
　trichocyte development in, 33—35
　unattached, 48
Neogoniolithon notarisii, trichocyte development in, 34
Neosolenopora armoricana, 174
New Zealand, and phytogeography of coralline algae, 120, 133, 144
Northern Bellairs Reef, 153
Norway
　Foslie Herbarium, 182
　phytogeography of coralline algae, 144
Nuclear tests, and uplifted coralline shores, 140
Nuclei, 37
　cellular fusion, 19, 20
　characteristics of, 20
　in female conceptacles, 101
　in relict rhodoliths, 154
　starch grain formation, 22
　tetrasporangial, 92

O

Oahu Island, fringing reef complex of, 162—163
Oceans, see also Atlantic Ocean, Pacific Ocean
　role of coralline algae in, 8—9
Organic matrix theory, 114

P

Pacific Ocean
　atolls of, 159
　　northeastern, 129
　　northwestern, 127—129
　　Porolithon in, 163—164
　reef production in, 141, 162—163
Paleozoic Era, 174
Parachaetetes, 174
Parrot fishes, grazing by, 161
"Pavement nullipore", 164
Penicillus, calcifying algae in, 111
Perithallia, 40
　in taxonomic system, 185
　structure of, 4
Peyssonellia, aragonite-producing, 111

pH levels
　and calcification process, 115—116
　and calcite production, 156
Phosphate
　effect on morphology of, 65
　in calcite crystallization, 115
　intolerance of, 143
Photosynthesis, and calcification process, 114—116
Phycologia, trichocytes in, 33
Phyllospadix, biotic interactions of, 146
Phymatolithon
　classification of, 222
　conceptacles, 82, 84
　epithallia of, 29, 31
　fossil, 173
　hypothallus in, 39, 40
　phytogeography of, 121, 122, 124, 125, 128
　thick crusts of, 47
　unattached, 48
Phymatolithon calcareum
　cell walls in, 14
　in commercially collected marl, 146
　water motion, 141
Phymatolithon laevigatum, cellular structures in, 22
Phymatolithon lenormandii
　cell fusion in, 19
　cellular structures in, 22
Phymatolithon polymorphum, salinity levels tolerated by, 143
Phymatolithon rugulosum, trichocytes in, 35
"Physiological low tide level", 159
Phytogeography
　Atlantic Ocean
　　northeastern, 124—127
　　northwestern, 119—124
　　below 30° south latitude, 133
　biotic reefs, 160
　Pacific Ocean
　　northeastern, 129
　　northwestern, 127—129
　paleobotany, 173
　tropical regions, 129—133
　12 major regions, 119-121
Pit bodies, cellular connections in, 16
Pit plug, 16, 17, 29
Plastids, 20, 21, 37, 97
Pleistocene platforms, 159
Polyporolithon, spiphytic species in, 52
Polyporolithon parcum, epiphytic existence of, 49
Population ecology, and light intensities, see also Light, 139
"Pore cells", 91
Porolithon
　classification of, 222
　in Pacific, 163
　in ridge formation, 168, 169
　magnesium content in, 112
　measurement of, 153
　megacell development in, 34, 35

phytogeography of, 119, 122, 128
productivity of, 152
protuberance formation in, 48
reef building of, 160—162
thick crusts of, 47
trichocyte development in, 33—35
"*Porolithon*-cap", 169
Porolithon gardineri
  thick crusts of, 48
  trichocyte development in, 34
Porolithon onkodes
  adaptive ability of, 138—139
  fringing reef complex, 164
  grazing, 161
  reef production of, 163
  sporeling development in, 24, 25
  thick crusts of, 48
  trichocyte development in, 34
  water motion, 140
Porolithon pachydermum
  and grazing activity, 161
  productivity of, 151
  reef building of, 169
Porolithon sonorense, 2
Primary meristems, 27
Primary pit-connections, 16
  arching lines formed by, 58
  spore germination, 23
Production
  growth rates, see also Growth rates, 149—150
  inorganic, 152—153
  organic, 151—152
Productivity
  and water motion, 141
  measurement of, 138
Prymnesiophyceae, calcite-producing, 111
Pseudodichotomous branching, in meristematic cell, 28
Pseudolithophyllum, phytogeography of, 122
Ptilothamnionopsis, biotic interactions of, 146

R

Red Sea, and phytogeography of coralline algae, 120, 133
Reef(s)
  bank-barrier, 168
  basic structure of, 195—161
  Bermuda, 167
  Caribbean, 164—170
  Caribbean fringing, 153
  cup, 166
  fringing, 153, 168
  grazing and coralline development, 161
  Hawaii, 160
  Indo-Pacific, 162—163
  morphogenesis of, 166
  Porolithon, 163—164
  tropical, 151
Reef flat, 160
Repopulation, 150

Reproduction
  bisporangia, 91—92
  carpogonia, 98—107
  carposporophytes, 101—106
  conceptacles, see also Conceptacles, 79—84
  life histories, 106—108
  male conceptacles, 92—97
  spermatangia, 97—98
  tetrasporangia, 84—92
  vegetative, 79
Reproductive initials, 91
Rhipocephalus, calcifying algae in, 111
Rhodoliths
  characteristics of, 154
  growth rates for, 149
  phytogeography of, 131—132
  structural features of, 2, 48
  water motion, 141—142
Rhodophyta (phylum), 12
Ribbon corallines, 46

S

St. Croix
  ecological relationships in reefs of, 168
  phytogeography of coralline algae, 138, 140
  reef production, 164
  reef system of, 168
Salinity, and growth rates, 142
San Clemente Island
  intertidal community of, 152
  phytogeography of coralline algae, 145
  sewage studies at, 143
Scanning electron microscopy
  of coralline surfaces, 188
  of thallus surfaces, 32
Schmitziella
  classification of, 222
  parasitic species of, 55
  phytogeography of, 125
Schmitziella cladophorae
  lack of calcium carbonate in, 16
  parasitic existence of, 55
Schmitziella endophloea
  lack of calcium carbonate in, 16
  parasitic existence of, 55
Schmitzielloideae (subfamily), 10
Sea level, and reef morphogenesis, 166
Seasonality
  and reproduction, 107
  of growth and reproduction, 136
Seawater ingredients, and growth rates, 142
Secondary pit-connections, 16
  direct, 17—18
  in Lithophylloideae, 41
Secondary tissues, 28
Serraticardia
  branching in, 66
  classification of, 222—223
  conceptacles of, 83
  effect of seasons on, 137

phytogeography of, 123, 130, 131
Serraticardia maxima, calcification in, 112
Sewage
  effect on coralline algae, 8
  effect on growth rates, 143
Shading, and reef production, see also Light, 164
Solenopora, 174
Solenoporaceae (extinct family), 10, 173—175
Spermatangia, 97—98
  in male conceptacles, 97
  nuclear conditions for, 99
Spermatangial initials, 87, 93
Spore(s)
  germination of, 22
    Amphiroa type, 24
    Corallina type, 24
    Dumontia type, 22—23
    Naccaria type, 22, 26
  size of, 23, 38
  substrates, 135
Sporeling(s)
  cell division in, 24
  development of
    Lithophyllum mode, 25, 26
    Lithothamnium mode of, 25
    Neogoniolithon mode of, 25, 26
  frond initiation, 65
Spore walls, structure of, 22
Sporolithon, 44
  cellular connections in, 20
  classification of, 223
  conceptacles, 80—81, 91
  effect of light on, 139
  epithallia of, 31
  phytogeography of, 122
  reef building of, 160—162
  reproductive initials in, 91
  taxonomic system for, 182
  thick crusts of, 47
Sporolithon erythraeum, habitat of, 162
Squamariaceae (family), 173
Staining, of living algae, 153
Staining bodies, 37
Stalk cells, 92
Starch grains, 21, 37
Substrate(s)
  coralline algae as, 9
  in taxonomic system, 184, 185
  spores and, 135
Subterraniphyllum, 177
Succession, 150—151
Supporting cells
  and fusion cells, 103
  in carpogonia, 99
Surfaces, see also Crusts
  electron microscopy of, 188
Surgeonfishes, grazing by, 161
Synarthrophyton
  cellular connections in, 20
  classification of, 223
  tetrasporangial conceptacles in, 89
Synarthrophyton patena

conceptacle formation in, 91
epiphytic existence of, 49, 51
gametophytes, 84
margin of, 51
Synchronous divisions, in terminal meristems, 27

T

Tabular keys, 183, 185
  for Mastophoroideae, 186
  for Melobesioideae, 185
  to Corallinoideae, 186—187
Taxonomy
  adaptations in coralline algae, 188—191
  current schemes of classification, 187—188
  history of, 179—183
  recognizing genera, 183
  suprageneric classification, 188
  tabular keys for, 183, 185
Temperature, see als Environment, 132
  and calcite production, 156
  and growth rates, 137
Tenarea
  classification of, 223
  diagnostic features of, 43
  effect of light on, 138
  phytogeography of, 125, 128
Tenarea tessellatum, diagnostic features of, 44
Tenarea tortuosa, and water motion, 140
Terminal meristems, 27
Tetrasporangia
  development of, 91—92
  structure of, 7
Tetraspores, 23, 24, 135—136
Tides, see also Water motion
  "physiological low tide level", 159
  ridge formation, 169
Transfer tubes, 102
Trichocytes
  as diagnostic features, 35
  development of, 32—34
  intercalary, 34
  terminal, 34
Trichogynes, 99
Tropical regions
  calcium carbonate in, 155
  phytogeography of, 129—133
  reef growth in, 152—153
  temperature related distributions, 137

U

Udotea, calcifying algae in, 111
Uniporate development, 84

V

Vacuoles, characteristics of, 21
Vegetative tissue, diagrams of, 6
Vermifuge, coralline algae as, 147

W

Water motion, see also Environment
    and adaptation of coralline algae, 190
    and growth rates, 140—141
    and reef morphogenesis, 166
    and reef production, 164
    and ridge formation, 169
    effect on coralline algae, 8
Water temperature, see Temperature
Waves, see Water motion
West Indies, algal ridge areas of, 164—166
Winds, and reef production, 164

X

X-ray diffraction, for quantifying calcification, 112

Y

Yamadaea
    adaptation of, 190
    branching in, 66
    classification of, 223
    conceptacles of, 83
    fronds of, 57
    phytogeography of, 123, 130, 131
Yamadaea melobesioides, growth forms in, 5

Z

Zoophytes, 10
Zostera marina L., with thin crust, 44